Macrophysics and geometry

From Einstein's unified field theory to cosmology

Macrophysics and geometry

From Einstein's unified field theory to cosmology

A. H. KLOTZ

Reader in Applied Mathematics, University of Sydney

Cambridge University Press

Cambridge

London   New York   New Rochelle

Melbourne   Sydney

7282-9023

Published by the Press Syndicate of the University of Cambridge
The Pitt Building, Trumpington Street, Cambridge CB2 1RP
32 East 57th Street, New York, NY 10022, USA
296 Beaconsfield Parade, Middle Park, Melbourne 3206, Australia

First published 1982

Printed in Great Britain at the University Press, Cambridge

British Library cataloguing in publication data
Klotz, A. H.
Macrophysics and geometry.
1. Fields, Algebraic
I. Title
512′.74   QA247
ISBN   0   521   23938   9

Library of Congress catalogue card number: 81-3849

*To my Wife, Wendy Ann
and to the Memory of a Genius*

Contents

		Page
Preface		ix
Abbreviations and notation		xii

1 Unified field theory and general relativity   **1**

1	The aim of a unified field theory of macrophysics	1
2	Remarks on theoretical and empirical knowledge	3
3	Some criteria of a good theory: philosophy and physics	6
4	A critique of general relativity	8
5	Development of GFT	13

2 Field equations of the generalised field theory   **18**

1	Principles of GFT: charge cojugation and Hermitian symmetry	18
2	Mathematical formulation of the principle of Hermitian symmetry	22
3	Variational parameters and the field equations	26
4	Why weak field equations?	29
5	Uniqueness and gauge invariance of WFE	31
6	The identities and conservation laws of GFT	32
7	The energy–momentum tensor in GFT	36

3 The Riemannian background of the generalised field theory   **39**

1	Bifurcation of geometry and physics: the weak principle of geometrisation	39
2	The conditions on a metric	41
3	The local Poincaré group	43
4	The metric hypothesis of GFT	45
5	The weak principle of geometrisation revisited	48
6	Significance of the metric hypothesis	50
7	A plan of action	51

4 The problem of motion   **54**

1	An outline of the Einstein–Infeld–Hoffman method	54
2	On Newtonian approximation and field strength	59
3	Equations of motion in GFT	61
4	A new law of motion	65

5 Conditions for an identification of the electromagnetic
field tensor | 67
6 The tensors $w_{\mu\nu}$ and $R_{[\mu\nu]}$ $(\tilde{\Gamma})$ | 69
7 A non-Maxwellian electrodynamics | 72
8 Equations of motion in a cylindrical field | 74

5 Solution of the field equations | **76**

1 The Tonnelat affine connection | 76
2 On geometrical symmetry | 79
3 On universal symmetry of Gregory | 84
4 The Vanstone solution | 86
5 The static, spherically symmetric metric of GFT | 90
6 The Coulomb law | 94

**6 The cosmological model as a consequence of the generalised
field theory** | **98**
1 Local and cosmic coordinates | 98
2 The light tracks (null geodesics) | 100
3 Isotropy and expansion of the world | 103
4 An oscillating universe? | 107
5 Geometry at infinity | 108
6 Time and distance | 110
7 Distribution of matter in the universe | 113
8 Volume and mass of the world | 115
9 Cosmic observables | 116
10 Inside $2m$ | 119
11 Cosmology without a cosmological principle | 119
12 Charge and mass | 120

7 Generalised field theory and microphysics | **122**

1 Spinors in the nonsymmetric theory | 122
2 Spinor analysis | 124
3 Some relations between 'tilded' and untilded tensors | 127
4 A comment on the minimum coupling hypothesis | 128
5 Flat space–time in GFT | 130
6 Dirac equations and the metric hypothesis | 131

8 An outlook | **135**

1 Generalised field theory as a completion of relativity | 135
2 Objections to the nonsymmetric theory | 138
3 The possibility of an empirical test of GFT | 141
4 Unsolved problems | 143

Bibliography | 147
Index | 150

Preface

It is commonly held that Einstein, having spent the last thirty years of his life in a search for a unified theory of all physics, failed to find one. And perhaps it is true that he failed to create a comprehensive account of all aspects of physical reality. Indeed, such an account seems to transcend our present day mathematical techniques or, for that matter, our understanding of the way the world – in all its aspects, of macrophysical continuity and of microphysical discreetness – appears to be constructed. But there are only two fields which are known to be effective on a laboratory or cosmic scale of experience and which conform to the concept of an indefinitely divisible and almost everywhere (a.e.) continuous space–time manifold or substratum in which physical events take place.

What does not seem to be appreciated is that Einstein did succeed in constructing a theory which gives a unified description of these macrophysical, or long range, fields of electromagnetism and gravitation. At any rate, Einstein did give the basic ideas on which such a theory can be founded. This monograph, which makes coherent the record of articles published over the last decade or so by the author and his students, is a presentation of the theory as it emerged from the fundamental concepts established by Einstein and his collaborators.

It is, in fact, a completion of the cycle of thought began by special relativity and continued through the 1915 theory of the gravitation field. It is the third theory of relativity which will be shown to await today only that which is the final test of any theory claiming to be a description of physical reality, namely, an empirical confirmation or rejection. In the latter case, failure of the theory will constitute a demonstration that for an incomprehensible reason, namely the equivalence principle of general relativity, Nature treats the two macrophysical fields, one in a very different way from the other; and that there is something unique about gravity. Such a demonstration cannot be logically arrived at from within the general relativistic reasoning.

In the unified field theory, the principle of equivalence is replaced by a certain formulation of the fact that the laws of physics are insensitive to a conjugation of the electric charge. This formulation, the principle of Hermitian symmetry, contains the general relativistic equivalence, but the converse is demonstrably untrue.

The theory of Einstein, or rather of Einstein and Straus, which for several reasons I shall explain is the starting point of my investigation, encountered severe theoretical objections before it could be developed to a stage at which it could be tested empirically. The most damaging of these was the apparent impossibility of deriving the equations of motion of a charged test particle from the field equations. Another objection, though never explicitly stated, was the equally apparent non-uniqueness of the field equations themselves. The nonsymmetric theory therefore seemed to be just another of the multitudinous proposals for a unified field theory, and moreover, one least likely to bear any resemblance to reality because of its formal faults. The electromagnetic potential required to yield a Lorentz force was soon discovered but no reason could be given for its peculiar form.

Why should there have been so many suggestions of how a comprehensive theory of electromagnetism and gravitation was to be constructed? This is difficult to answer without indulging in a psychological speculation. I think that it was basically due to a misunderstanding of what exactly was required. And, emphatically, it is not just to find a structure with a covariant vector to be glibly interpreted as the electromagnetic potential. A theory of physics ought to begin in empirical evidence, although the very theory which Einstein sought to supplant, his own eminently successful account of gravitation contained in general relativity, did not! The latter has its origin in a conceptual dissatisfaction with the restrictions of the special theory. Similarly, there are powerful logical reasons why a unified field theory should give a better description of reality than the theory of the gravitational field in a curved space–time.

Because of this background, I begin the present monograph with a brief critique of other than the nonsymmetric proposals for a unified field theory and with an attempt to establish the claim of the nonsymmetric theory to represent the correct approach to the problem from a philosophical point of view. The theory I am discussing constitutes a reformulation of Einstein's original concept, forced by problems not considered by him. Consequently, rather than showing immediately the lack of foundation of the specific objections raised against it, I shall describe first the complete structure of the unified field with particular emphasis on the principles from which it follows. The solution of the problem of motion will then appear naturally as part of the theory.

It will also follow that the unified field theory is unlikely to be testable 'in the laboratory'. Indeed, one must look to its 'large scale' consequences to find any observable deviation from the prediction of general relativity or of classical electromagnetism. Here, a surprise awaits us. It seems that the only reasonable interpretation of the mathematical structure of the theory is that it should yield, in a certain sense to be discussed in detail, a unique cosmological model. I conclude the monograph, therefore, with an account, as far as it has been investigated, of the major features of this model.

My sincere gratitude is due to Ms Jane Louise Markham for invaluable help with typing the manuscript and designing the jacket.

Alex Klotz

Sydney 1981

Abbreviations and notation

SR   Special relativity
GR   General relativity
GFT   Generalised field theory: in particular, the nonsymmetric unified field theory of gravitation and electromagnetism first proposed by Einstein and Straus in 1946, which is the main topic of this monograph
WFE   Weak field equations (of Einstein and Straus)
SFE   Strong field equations (Einstein, 1945)
PC   General principle of covariance
WPG   Weak principle of geometrisation
PR   Principle of reducibility
PHS   Principle of Hermitian symmetry, double transposition or of charge conjugation (equivalent terminologies): the central principle of GFT, replacing the principle of equivalence of general relativity

Einstein's summation convention over repeated indices is used consistently unless it is explicitly stated otherwise.

Greek indices α, $\beta, \ldots, \mu, \nu, \ldots$ go from 0 to 3: with x^0 time-like and x^1, x^2, x^3 space-like.

Latin indices (lower case) $i, j, \ldots$ go from 1 to 3.

Capital Latin indices (spinor indices) $A, B, \ldots$ or (complex conjugates) $\dot{A}, \dot{B}, \ldots$ go from 1 to 2.

V_4: a Riemannian four-dimensional space–time of signature -2

$\quad x^\mu = (x^0, x^1, x^2, x^3)$:   a coordinate system in the geometrical manifold.

$\quad$ Metric of V_4: $\mathrm{d}s^2 = g_{\mu\nu}\mathrm{d}x^\mu\mathrm{d}x^\nu$ (with nonsingular, symmetric $g_{\mu\nu}$, but note GFT)

$\quad$ or $\quad \mathrm{d}s^2 = a_{\mu\nu}\mathrm{d}x^\mu\mathrm{d}x^\nu$ (with nonsingular, always symmetric $a_{\mu\nu}$ metric tensor)

$\left\{\begin{matrix}\lambda\\\mu\nu\end{matrix}\right\}_g, \left\{\begin{matrix}\lambda\\\mu\nu\end{matrix}\right\}_a$: Christoffel brackets (of the second kind, the affine connection of V_4) constructed from $g_{\mu\nu}$, $a_{\mu\nu}$ respectively.

$A_{(\mu\nu)} = \frac{1}{2}(A_{\mu\nu} + A_{\nu\mu})$: the symmetric part of a two-index (covariant) quantity.

$A_{[\mu\nu]} = \frac{1}{2}(A_{\mu\nu} - A_{\nu\mu})$: the skewsymmetric part of a two-index (covariant) quantity.

For a many-index quantity:

$$A^{\cdots}_{\cdots(\lambda/\mu\nu\ldots/\sigma)\ldots} = \frac{1}{2}(A^{\cdots}_{\cdots\lambda\mu\nu\ldots\sigma\ldots} + A^{\cdots}_{\cdots\sigma\mu\nu\ldots\lambda\ldots});$$

and likewise for the square bracket.

Similarly for contravariant indices:

$$A^{(\mu\nu)}, A^{[\mu\nu]}, A^{\cdots(\lambda/\mu\nu\ldots/\sigma)\cdots}_{\cdots}, A^{\cdots[\lambda/\mu\nu\ldots/\sigma]\cdots}_{\cdots}.$$

$A_{[\lambda\mu\nu]} = \frac{1}{2}(A_{\lambda\mu\nu} + A_{\mu\nu\lambda} + A_{\nu\lambda\mu})$, (and $A^{[\lambda\mu\nu]}$ similarly)

$\Gamma^{\lambda}_{\mu\nu}$: an affine connection (usually nonsymmetric, $= \Gamma^{\lambda}_{(\mu\nu)} + \Gamma^{\lambda}_{[\mu\nu]}$)

$\Gamma_{\mu} = \Gamma^{\sigma}_{[\mu\sigma]} = -\Gamma^{\sigma}_{[\sigma\mu]}$

$\tilde{\Gamma}^{\lambda}_{\mu\nu} = \Gamma^{\lambda}_{\mu\nu} + \frac{2}{3}\delta^{\lambda}_{\mu}\Gamma_{\nu}$: Schrödinger's affine connection ($\tilde{\Gamma}_{\mu} \equiv 0$) ($\delta^{\lambda}_{\mu}$: Kronecker delta)

$\varepsilon_{\alpha\beta\gamma\delta}(\varepsilon^{\alpha\beta\gamma\delta})$: Levi-Civita permutation symbols (tensor densities)

In GFT, $g_{\mu\nu}$ is nonsymmetric but still nonsingular:

$$g_{\mu\nu} = g_{(\mu\nu)} + g_{[\mu\nu]},$$

and denotes a physical field tensor (not necessarily identical with 'known' physical fields).

For any covariant $((0,2))$ tensor $h_{\mu\nu}$, symmetric or nonsymmetric:

$h = $ determinant $(h_{\mu\nu})$.

Gothic letters denote tensor densities; for example:

$$\mathfrak{g}^{\mu\nu} = \sqrt{(-g)}g^{\mu\nu}; \quad \mathfrak{g}_{\mu\nu} = (1/\sqrt{(-g)})g_{\mu\nu}.$$

A comma followed by a (covariant) index denotes the partial derivative with respect to x^{μ}

$$h^{\lambda}{}_{,\mu} = \frac{\partial h^{\lambda}}{\partial x^{\mu}}, \quad h_{\lambda,\mu\nu} = \frac{\partial^2 h_{\lambda}}{\partial x^{\mu}\partial x^{\nu}} \text{ etc.}$$

(also $\partial_{\mu}h^{\lambda}$ etc. if convenient).

Semicolons denote covariant derivatives (also operators such as D_{λ} or ∇_{λ} if convenient, especially in chapter 7) with subscripts $+$, $-$ indicating order of indices in covariant derivatives with respect to a nonsymmetric affine connection; subscript 0 denoting covariant differentiation with respect to the symmetric part of a nonsymmetric connection; this is explained in the text but as examples we have

$$h^{\lambda}_{+;\mu} = h^{\lambda}_{,\mu} + \Gamma^{\lambda}_{\sigma\mu}h^{\sigma} \quad (h_{\lambda;\mu} = h_{\lambda,\mu} - \Gamma^{\sigma}_{\lambda\mu}h_{\sigma}),$$

$$h^{\lambda}_{-;\mu} = h^{\lambda}_{,\mu} + \Gamma^{\lambda}_{\mu\sigma}h^{\sigma},$$

$$h^{0\,\lambda}_{\;\;;\mu} = h^{\lambda}_{,\mu} + \Gamma^{\lambda}_{(\sigma\mu)}h^{\sigma},$$

$$g_{\mu\nu;\lambda}_{+\,-} = g_{\mu\nu,\lambda} - \Gamma^{\sigma}_{\mu\lambda}g_{\sigma\nu} - \Gamma^{\sigma}_{\lambda\nu}g_{\mu\sigma} \text{ etc.}$$

Riemann–Christoffel tensor:

$$R^{\lambda}_{\mu\nu\sigma} = -\Gamma^{\lambda}_{\mu\nu,\sigma} + \Gamma^{\lambda}_{\mu\sigma,\nu} + \Gamma^{\rho}_{\mu\sigma}\Gamma^{\lambda}_{\rho\nu} - \Gamma^{\rho}_{\mu\nu}\Gamma^{\lambda}_{\rho\sigma}.$$

Ricci tensor:

$$R_{\mu\nu} = R^{\sigma}_{\mu\nu\sigma} = -\Gamma^{\sigma}_{\mu\nu,\sigma} + \Gamma^{\sigma}_{\mu\sigma,\nu} + \Gamma^{\rho}_{\mu\sigma}\Gamma^{\sigma}_{\rho\nu}$$
$$- \Gamma^{\rho}_{\mu\nu}\Gamma^{\sigma}_{\rho\sigma}.$$

Einstein tensor (in a V_4 with the metric tensor $a_{\mu\nu}$):

$$G_{\mu\nu} = R_{\mu\nu} - \tfrac{1}{2}a_{\mu\nu}(a^{\alpha\beta}R_{\alpha\beta} - 2\Lambda).$$

Λ: the 'cosmological constant'

1 Unified field theory and general relativity

1. The aim of a unified field theory of macrophysics

The aim of the nonsymmetric unified field theory which is the subject matter of this monograph is to explain the limitations of general relativity. Only *a posteriori*, as a consequence of interpretation, does the theory become a comprehensive description of the gravitational and electromagnetic fields in interaction with each other. The term 'unified field theory' today covers a wide variety of concepts and I shall refer in the sequel to the particular theory initiated in 1946 by Einstein and Straus as the generalised field theory (GFT). I shall show the GFT successfully fulfils the above aims provided it is correctly interpreted in a way unknown to its original authors. Whether it is a tenable theory of physics depends now only on the final test of validity of any theory, namely a comparison with empirical evidence. But it is possible to derive from GFT consequences and results which are, at least in principle, so testable, and the theory makes verifiable predictions.

GFT is a completion of relativistic thinking through the respective theories of special and general relativity. It is then not surprising that it should turn out to be a unified account of the two fields, of electromagnetism and gravitation, unquestionably known to macrophysics. Einstein's special relativity (SR) was originally designed to handle ambiguities arising from Maxwell's synthesis of the electromagnetic field theory. It was, of course, his genius to recognise that the idea of observer equivalence, and therefore of the transformation invariance of laws, was applicable to all branches of physics and in particular to mechanics which enabled the theory to become so well established. As a result, however, SR became a theory of communication between human observers and then the restriction of equivalence to observers in a state of uniform relative motion appeared incomprehensible except as a first approximation to reality. SR also is not a field theory and, in spite of its success with electromagnetism, it is unable to describe correctly the effect of gravitation. Einstein overcame these shortcomings in general relativity (GR), but the ideas on which this theory was based promised a far wider applicability than the area of the pure gravitational field in which the theory met its greatest, if not its only empirical success. The group of allowable transformations was extended from the restricted Lorentz group to a general Lie group of continuous, and a.e. C^∞ transformations (i.e. with an a.e. nonzero Jacobian). Thus observers in arbitrary relative motion became equivalent. Also, the concept of geometrisation of physics was introduced, only hinted upon in SR through the four-dimensionality of a Minkowski space–time as the substratum in which physical events take place. But the principle of equivalence, which was perhaps inevitable when GR was proposed, strongly identifies a Riemannian space–time (V_4) with the gravitational field. The latter is the only thing that is geometrised by the theory. This curious limitation provided the original motive for the search for a unified field theory.

It was an aesthetic reason because, unlike the case of SR which after all failed to predict correctly the amount by which light is deflected in a gravitational field or to account for the amount by which the perihelion of a planet is shifted, there does not seem to be any empirical evidence which would force the search. It is simply that we do not have unambiguous evidence of a sufficiently strong and permanent electromagnetic field to observe, and powerful gravitational fields appear to be too remote for the effect of any possible interaction, and hence a possible failure of GR, to be detectable.

The aesthetic motive was also responsible for the multiplicity of proposals for a unified field theory. With the possible exception of the theory due to Weyl (which was also the first of these attempts if we except the ideas of G. Mie which predated general relativity), and the five- or more dimensional theories which are inexplicable, all of the other proposals started from what is virtually a wrong point. There are two reasons for this, both stemming from the above purely metaphysical but irresponsible motive. The first is that the theories began with the aim of unifying electromagnetism and gravitation as already understood; and the second, and this applies also to the five-dimensional theories that, following the pattern of general relativity, they sought to postulate a new geometry as the model of the combined field. Both can be done in many ways as experience has shown, but in the absence of physically meaningful principles on which the comprehensive theory could be founded and from which it could be derived, they mostly led to a mnemonic device for the derivation of the known equations. Thus, they were unable to predict an interaction of the macroscopic fields and were untestable.

This brings me to the current unified field theories of the type considered by Salam and Weinberg which at last appear to be meeting with some degree of empirical success. They are essentially attempts to unify quantised fields relying on the idea that these should merge at suitably high energy. Conceptually, they depend on the fact that physics is described at a fundamental level by gauge theories, and select this or that, SU(3) × SU(2), SU(8) and so on, symmetry group as the hypothetical starting point. This is geometrisation of a very different kind than that envisaged by Einstein. The resulting theories, whatever their value as models of an aspect of reality, cannot throw any light on the problem of why general relativity should be such a good theory of gravitation but of little else. They are basically model dependent, whereas what is needed is a model independent theory, encompassing GR, and like the latter derivable from physically meaningful assumptions. GFT is such a theory.

It is essentially a macroscopic theory in which events occur in a continuous, indefinitely subdivisible space–time manifold. Its field equations are differential equations whose dependent variables are thought of as directly measurable physical quantities. It may be a weakness but it is also a strength of GFT that it perpetuates the conceptual dichotomy between quantum mechanics and macrophysics.

It unifies the fields known on the laboratory, astronomical and cosmic scale of experience, which are just the electromagnetic and gravitational fields, although it contains an indication of its meeting point with the quantum level where the theories of Weinberg and Salam properly take over. All this, however, is achieved through a reformulation of GFT unsuspected by Einstein.

2. Remarks on theoretical and empirical knowledge

The origin of GFT lies in a critique of general relativity on conceptual rather than empirical grounds and this implies a certain philosophical attitude. Now philosophical ideas play a part in the formulation of the hypotheses from which a theory (of physics) is derived especially if there is little experimental evidence requiring explanation. This is the case not only with GFT but also with GR since strong gravitational fields are difficult to reproduce on the laboratory scale and we must rely on observations of distant events which are often interpreted by assuming the theory they are meant to test.

In order to be acceptable a theory must not only allow a comparison against known evidence. It should also forecast hitherto unobserved phenomena or at least explain known empirical facts which are inexplicable on the basis of theories which it aims to replace. If it achieves only the latter, its uniqueness remains questionable. In the former case, the theory is falsifiable in Popper's sense provided that evidence has been correctly interpreted. This condition weakens Popper's criterion considerably because it leaves open the question of what a correct interpretation may mean. Only very primitive theories can be falsified by a single contrary instance. It is also only then that falsifiability can be strictly deduced from the simple fact that a theory can not be verified either. All that can be said is that in the instances looked at, it has (for 'verification') predicted an outcome which was found to occur.

The subject matter of physics is the material world which, if physics is to be possible, must be assumed to be knowable by means of observation and experiment. It is also usually assumed that this world, or at least aspects of it, can be described. Reality is a synonym for the material world whose existence outside the imagination of its human observer is postulated as the field of physical interest. However, observers themselves belong to the world although they transcend it by reasoning about it and thus by creating physics. Perhaps this is why mathematics, apparently an abstract creation of the mind, is applicable. Or, maybe, mathematics is the language of Nature in a more concrete sense. We do not know.

The material world is by definition unique. If another one existed, with another physics describing relationships within it but without an interaction with the first, it would be of no concern to us as part of the former. If there was an interaction between them, they would not be distinct. One physics, defined as the set of all relationships and phenomena, would describe both worlds.

This does not mean that we must necessarily expect a single mathematical model for the whole of reality. It only follows (from its uniqueness) that any theory of physics must in some sense correspond to empirical knowledge. I mean by theory a mathematical model of reality or of an identifiable aspect of it together with all the deductions which may follow. Empirical knowledge is the totality of what we may learn about reality through experiment, observation and a classification of data.

A theory must be modified or rejected if it can not be reconciled with what is empirically known. However, it must always be remembered that the results of observation themselves are subject to misunderstanding and even the conclusions of a mathematical deduction can be misapplied leading to an apparent downfall of a perfectly tenable theory. The case of the problem of motion in GFT illustrates the latter. Actually, if a theory is well established, that is if there is a wealth of empirical evidence to support it, we expect that it only needs a modification if it is found that its area of applicability is limited. SR has not falsified Newtonian mechanics but is a better approximation to reality. Similarly it has not been replaced by GR, although GR gives a better description of the gravitational field. We can assert that the domain of applicability of GR is unknown but this is only a partial answer to why GFT is necessary.

In general, not only a theory and its hypotheses, but also interpretations of empirical evidence must be taken into account in considering the problem of verification. This is particularly important when pure experimenting is impossible. An obvious example of this is the science of astronomy. Experiment consists, in effect, of asking questions of nature in a manner in which it can supply answers. It follows that even when experimenting is possible it is necessary to consider whether the right question has been asked: is a given experiment allowed? In particular, does it belong to what is being investigated?

Popper's analysis of the validity of theories seems to ignore these wider aspects of the relationship between empirical and theoretical knowledge. And if they are not considered, science becomes a game with no rules. However, even apart from this there is another point which should be noted and which refers to the relationship between successive theories of the same aspect of reality.

A distinction should be drawn between the logical content of a theory and what may be described as its didactic promise. This is the extent to which the aims of a theory are realised within its actual area of applicability as it may be found. Because of its assumptions, a theory may well be wider than that part of it which has been verified. In principle, domains of applicability and of verifiability should coincide but in practice, the latter grows with every new experiment or observation unambiguously referring to the theory in question. For example, empirical evidence in GR renders it an excellent model of the gravitational field but little else. In particular, there is no evidence for or against GR with matter even when the energy–momentum tensor is purely electromagnetic (Einstein–Maxwell theory).

Is it possible to talk meaningfully in such circumstances about validity of the theory? An empirical test is certainly a necessary criterion for its validity but it is not sufficient. Indeed, it can not be sufficient in general, if the theory is sufficiently comprehensive, that is if it can be regarded as extending our understanding of Nature usually by correlating concepts which were thought to be unrelated before the theory was proposed. GFT is a theory of this type.

3.   Some criteria of a good theory: philosophy and physics
        For a theory of physics to be classified as 'good', its conclusions and predictions can not be in conflict with empirical evidence. I have perhaps said enough already to raise the question of whether this is always enough. Let us consider this point a little further.

Special relativity can be said to develop naturally into the general theory. The argument for a macrophysical unified field theory will begin with the assertion that GR is inadequate as a fundamental theory of physics. I shall consider more technical aspects of this claim in the next section. I want to confine myself here to a somewhat formal problem raised by the structure of GR and illustrated by figure 1. Let region ɪ represent the domain of applicability of the principles of GR and region ɪᴠ, with shaded boundary, its gravitational restriction ('GR is a good theory of gravitation but little else'). Region ɪ may be the macrophysics or perhaps the whole of physics. If the latter, let the broken contour ɪɪ represent macrophysics. By the general relativistic principles of covariance and geometrisation, ɪɪ should coincide with

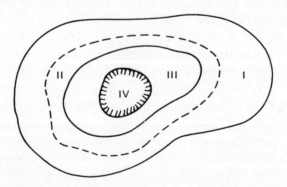

Figure 1

GR. Then the meaning of the shaded contour is intelligible only in the context of a theory more comprehensive than GR, such as GFT (iii).

This is basically a philosophical argument and it may not be out of place to comment briefly on the relation between physics and philosophy. The latter, or more precisely metaphysics, can not answer scientific questions which arise from empirical evidence. How often it went wrong is amply illustrated by such examples as biblical objections to the heliocentric hypothesis, proofs that there can be only seven planets in the Solar system, arguments that only Euclidean geometry is admissible, or rejection (without understanding) of the nature of the relativistic interpretation of time.

Nevertheless there exists a relation between philosophy and physics and it is well known that science can and does raise problems which it may be the task of philosophy to explore. I am interested here in the converse question of whether a philosophical argument and not just a general principle can be usefully employed in physics.

Philosophy is concerned with an understanding of the world which transcends that which can be learned from observation, experiment or from mathematical model building. It can not ignore the human agent because it is he or she who philosophises. It is also this agent who carries out experiments and tries to understand the world of Nature. And the identity of the scientist is not lost because the methods of science and philosophy may differ.

The theories of relativity, SR, GR and GFT, emphasise the part played by an observer who is necessarily human. Consequently, they have been beset from the start by questions concerning their philosophical background and meaning. Most of these questions were spurious and arose out of a confusion between physics and philosophy. For example, what relativity asserts is that physics is more intelligible if written in a four-dimensional language of Minkowski, Riemann or of a non-Riemannian geometry and not that the world is 'really' four-dimensional. However, the problem of what constitutes an intelligible theory belongs not only to physics but also to philosophy which can thus provide grounds for a critical evaluation of what we may wish to propose in physics.

A theory concerns our understanding of the world and it is on the meaning of this proposition, as well as on its logical and empirical consistency, that I base our discussion of the unified field. The three theories of relativity form a logical sequence in the speculative development of the theory of macrophysics. What is not often noticed is that they imply each other in the sense that the more restricted theory becomes intelligible only in the context of the more comprehensive one. It is not that it is thereby explained. What is explained is the meaning of the restriction. Within the context of the theory to which the restriction applies, it is arbitrary. The arbitrariness is overcome only in the context of the theory with a wider domain of applicability. The axioms of the theory contain also what I have called a 'didactic promise'. Thus, the fulfilment or otherwise of its stated aims can serve as a measure of the extent of its meaning. In this sense GR fails and GFT succeeds.

The foregoing remarks can be summarised as a principle of hierarchy of theories or perhaps as a principle of adequacy. GR is necessary not only to describe the gravitational field but also to explain the special relativistic restriction of observer-equivalence to observers in a state of uniform relative motion. A restriction such as this is purely arbitrary within the context of the theory to which it applies. Its full implication, and therefore also the adequacy of the restricted theory, can only be understood if the restriction itself is relaxed. This process invariably leads to a new theory which may or may not be a valid model of reality or of some of its aspects. If it is not valid, if it conflicts with observed data or if its predictions fail to correspond to empirical facts, the failure of the generalisation can be regarded as a good reason to suspect that the original restriction represents a fact of nature. But the proof requires caution. Sufficiently comprehensive theories are rarely falsified by isolated failures of their predictions especially when interpretation of data is uncertain. And physical theories must also be intelligible in the sense of being free, as far as possible, from arbitrary assumptions. Investigation of the extent to which basic hypotheses of a theory are arbitrary, leads to an ordered sequence of theories of consecutively higher, that is more comprehensive status.

SR, GR and GFT form such a sequence in the development of our understanding of macrophysics. Because of this it is sufficient to consider in more detail than hitherto the problems arising from general relativity. It will be a sufficient basis of GFT, the subject matter of this monograph, if these are found to be real enough.

4.   A critique of general relativity

General relativity is a good theory of the gravitational field according to all conceivable criteria of what a 'good theory' may mean. Its predictions have been confirmed empirically to a remarkable degree of accuracy. In comparison with other physical theories, and notably with quantum mechanics, the empirical evidence, and certainly experimental evidence, for GR may appear scanty but this does not diminish the certainty with which the above assertion can be made. The main reason is perhaps the spectacular nature of the confirmation of conclusions drawn from a theory which represents an enormous mathematical complexification of Newtonian theory notwithstanding its conceptual comprehensiveness and beauty. The major achievement of GR, of course, is the concept of identification of geometry and (an aspect of) physics.

Nevertheless, when talking about GR it is necessary to distinguish very carefully whether we mean the theory based on the field equations:

$$R_{\mu\nu} = 0 \tag{1}$$

of the pure gravitational field outside any sources which may give rise to it, or the theory

$$G_{\mu\nu} = -\kappa T_{\mu\nu} \tag{2}$$

whose field equations contain these sources explicitly (in the energy–momentum tensor $T_{\mu\nu}$). The vanishing of the Ricci tensor (equation (1)) follows from Einstein's recognition of the unity of gravitation and inertia (the weak principle of equivalence; the equality of gravitational and inertial mass being one of the most accurately attested facts of Nature). Similarly implied is the identification of the components of a Riemannian metric tensor, from which together with their first and second derivatives the Ricci tensor is constructed. Empirical confirmation of the theory followed from an analysis of the equations of motion of a test particle except for the gravitational red shift, an immediate consequence of the Schwarzschild solution. These, however, as geodesics of the V_4, are implied by the field equations.

On the other hand, equations (2) are written down because the Einstein tensor $G_{\mu\nu}$ is the unique tensor in a Riemannian geometry which is conserved and which satisfies the condition of linearity in the second derivatives of the metric tensor. And whether we are right or not, we believe that the energy–momentum tensor is always conserved. The field equations (2) are then a hypothetical prescription of how matter and any energy-carrying field determines geometry of the space–time and, *a fortiori*, the gravitational (or metric) field. Of course, any solution of equations (1) also determines both the latter. I shall refer to this codetermination as the strong principle of geometrisation. But equations (2) imply something else, and that is that gravity is a privileged force of Nature which alone describes the geometry (or is described by it). Everything else is bundled on the right hand side under the guise of energy–momentum and can only be a source of gravity (and of geometry).

The argument runs roughly as follows. We say that inertial mass, that is the measure of a dynamical response of a body to a distortion of space–time or force, is also the source of a gravitational field because of mass equivalence. Now, energy generates mass since by SR they are proportional. Hence any energy-carrying field generates gravitation. But the field itself is generated by a charge appropriate to it. Hence any charge-carrying field generates a gravitational field in much the same way as the inertial mass.

On the other hand, the field of a nongravitational charge is distinct from the gravitational field. This raises immediately the problem of whether it is legitimate to treat nongravitational fields in a different way from gravitational ones and whether or not they can affect the geometry as the latter does. All material bodies carry an inertial mass and attract each other through self-generated gravity unless something else, say an electric charge, is painted over them. By analogy with the electromagnetic field, we can say that as far as gravitational interaction is concerned, any two material bodies behave as if they were of opposite charge; simply because of the presence of the other. It may be that the existence of such a seemingly unique force against which no screening appears to be possible is a necessary prerequisite of an order in Nature and hence of physics. An electromagnetic field can be screened off mainly because of the dual sign of charge and the consequent possibility of electric dipoles. It is also the simplest of such fields, at least for electromagnetically isotropic media and for its Maxwellian model. Thus, the two macrophysical fields represent respectively the two fundamental modes of interaction in a continuous space–time. This has nothing to do with geometry but it would be a strange freak of Nature if one and not the other were to be identified with the geometrical structure of the world unless the whole concept of geometrisation was wrong and this seems unlikely in view of the empirical success of GR.

Apart from introducing geometrisation of physics, GR also removed from it the need to consider inertial reference systems. Levi-Civita, however, pointed out that this does not depend on the metric structure but on the displacement of the field, that is on the affine connection. The latter is a Christoffel bracket (Riemannian geometry) if it is symmetric and if the vector lengths are unchanged under the displacement (parallel or 'no change' transfer which has little to do with length invariance).

Either restriction represents an additional assumption and is clearly removable. The fact that geodesic equations can be derived from the field equations and can describe the motion of a free body or of a test particle (i.e. a body whose self-field and geometrical structure are negligible) in a gravitational field, is an argument for the privileged role of gravitation only if one adheres to stronger forms of the principle of equivalence than mass equality. For example (Misner, Thorne and Wheeler), one could claim that the laws of physics reduce in the vicinity of any point of the universal space–time to their special relativistic form. Indeed, a local Lorentz group generates GR but the above form of the principle smells of the occult.

As Einstein said: '... From this principle [of equivalence, irrespective of form], in conjunction with the knowledge that light exhibits a definite behaviour in empty space, it followed that the properties of the latter were to be represented by a symmetrical [Riemannian metric] tensor, $g_{\mu\nu}$. The principle of equivalence, however, does not give any clue as to what may be the more comprehensive mathematical structure on which to base the treatment of the total field ...'. This was not the only complaint of the great physicist against his own theory. Perhaps a stronger one was the fact that however inevitable the field equations (2) may have appeared when the theory was being set up, GR contains no recipe of how the energy–momentum tensor is to be written down in a given physical situation. Indeed it can only be constructed on the basis of 'prerelativistic' physics, that is of the view of Nature which takes no account of the inherent geometrisation. I shall return to the difficulty to which this leads in a moment, but the question which occurs first is: what is the total field?

As far as macrophysics is concerned, we know the electromagnetic and the gravitational fields. I have already suggested however, that the desire to reproduce the equations of general relativity and of Maxwell's electromagnetic theory was responsible for the multiplicity of schemes only. Einstein was well aware of this when he wrote:'... The only clue which can be drawn from experience is the vague perception that something like Maxwell's electromagnetic field has to be contained within the total field.' Something like Maxwell but not necessarily Maxwell...: only empirical evidence (and of course, interpretation) based on the consequences of a more general theory than GR can tell what the total field may be.

The same criticism applies to the Einstein–Maxwell theory,

$$G_{\mu\nu} = -\kappa E_{\mu\nu}, \tag{3}$$

where $E_{\mu\nu}$ is the Maxwell energy–stress–momentum tensor of the electromagnetic field. It presupposes the absolute validity of Maxwell's theory which must be only an approximation to the behaviour of electromagnetism especially in interaction with gravity. Is this interaction tractable?

There is a rule particularly in astronomy that no new physics should be proposed (to explain a phenomenon: e.g. Hubble expansion of the universe) if 'old' physics will do. Such a rule is essential but a rider must be added to it: provided the known theory is well established and well understood. But GR of the field equations (2) or (3) is neither.

So far we have considered only formal objections to GR: the nongeometrisation of anything other than gravitation; the non-applicability of the equivalence principle to the concept of total field; lack of theoretical indication of how the energy–momentum tensor is to be constructed; and earlier, the impossibility of accounting for its own limitations (of circumscribing its domain of applicability) from within. Let us now ask not whether there is any empirical evidence to support the field equations (2) as there is no direct evidence, but whether these equations are verifiable empirically at all. Because it is impossible to produce strong gravitational fields, wherein alone relativistic correction can become observable, any evidence must come from distant (almost always extragalactic) objects. However, this evidence is invariably interpreted by assuming GR and therefore can not be used to confirm it while avoiding a circular argument.

I shall confine myself here to simply raising the question, since a detailed analysis of verifiability of the equations (2) is outside the scope of this monograph. As far as GFT is concerned, it is enough that the question can be raised. Let us consider the Reissner–Nordström solution of the Einstein–Maxwell equation (3). The geometry of space–time can only be investigated and discovered by observing the motion of a test particle capable of interacting with all the fields present or supposed to be present. In the case of the Reissner–Nordström geometry, the test particle must be electrically charged. Then three questions arise: is the test particle a right one, is the source of the field really a stationary charge, and has the energy tensor been correctly constructed. All these, and possibly more, have to be independently answered by the experiment before the result of an observation can be asserted to confirm or falsify the theory. It seems that the third question, in particular, cannot be answered even in principle. The conclusion therefore appears to be inescapable that the field equations (2) cannot be empirically verified.

But a similar objection can be raised against almost any theory of physics. What can be done is to minimise the number of quantities which lead to empirical problems. In the case of the theory of relativity, it means attempting to write the field equations without $T_{\mu\nu}$. This in fact is the reason for the remarkable success of the field equations (1). If for some reason or other, they do not describe enough physics, if their domain of applicability is too narrow, then they must be generalised and extended. This does not mean that the new theory will not have energy–momentum but that $T_{\mu\nu}$ is to be calculated once the field equations are solved and not postulated before the solution is even attempted. Indeed, $T_{\mu\nu}$ can be calculated through something like the equations (2) (this is the case of GFT), and this will then imply that the space–time is always adequately described by a Riemannian model. The only question will be how the V_4 is to be determined. In GFT, the answer will imply a curious bifurcation of geometry and physics, a price to be paid for the comprehensive description of the macrophysical fields. Nature is not as simple as we may imagine.

5.   Development of GFT

GFT suffers, in company with all other proposals for a unified field theory of macrophysics, from working in an almost total empirical vacuum. In the circumstances, no matter how well founded the motivation for proposing the theory may be, the completed structure presents a somewhat stark aspect. The assumptions from which it is derived, in particular, can not escape a philosophical flavour, nor therefore being subject to dispute. Of course, any physical theory stands or falls by its ability to account for empirical facts but, with respect to a new proposal, it is hard to avoid the reaction: 'the bride is too lovely' (a free translation of 'die Breit ist zu schön'). The compulsion of a theory on the other hand is often enhanced by tracing its conceptual development.

GFT is no exception. Its full implications, especially as far as the implied relation between geometry and physics, could not be realised until most of the consequences were known.

I do not propose to outline here the whole history of the non-symmetric unified field theory but only to mention some of the crucial events which led to its final formulation. I have already referred in passing to the fact that of all the restrictions of Riemannian geometry (actually of general relativity as identified with a Riemannian model and therefore of the strong principle of geometrisation) the one which can be most easily discarded is the symmetry of the affine connection. In V_4 it arises from the assumption of closure of infinitesimal parallelograms, but this requirement can have little physical content. As early as the mid-twenties, that is before his momentous correspondence with Elie Cartan, Einstein considered identifying

$$\Gamma_\mu = \Gamma^\sigma_{[\mu\sigma]} \tag{1}$$

as the electromagnetic vector potential. This identification will be reintroduced in GFT. But Einstein could not go further then, for a variety of reasons though primarily because he did not have a physical basis for his theory.

Two versions of the nonsymmetric theory followed as soon as he arrived at a convincing principle, that of Hermitian symmetry or charge conjugation invariance. Hlavaty calls this an intuition of genius because Einstein did not have a suitable field tensor. Actually Hlavaty's Maxwell tensor was still wrong. The theory was shrouded in an interpretation incompatible with the basic principle and led to unanswerable objections. With a nonsymmetric (non-singular) tensor

$$g_{\mu\nu} = g_{(\mu\nu)} + g_{[\mu\nu]} \tag{2}$$

for which he retained a direct geometrical meaning (nonsymmetric 'metric') and a nonsymmetric affine connection

$$\Gamma^\lambda_{\mu\nu} = \Gamma^\lambda_{(\mu\nu)} + \Gamma^\lambda_{[\mu\nu]}, \tag{3}$$

Einstein proposed in 1945 the so called strong field equations (SFE). These are

$$\left.\begin{aligned}
g_{\mu\nu,\lambda} - \Gamma^\sigma_{\mu\lambda}g_{\sigma\nu} - \Gamma^\sigma_{\lambda\nu}g_{\mu\sigma} &= 0, \\
\Gamma_\mu &= 0, \\
R_{\mu\nu} &= 0.
\end{aligned}\right\} \tag{4}$$

Einstein thought that SFE implied four identities which would have made them derivable from a variational principle. This was not so except in a very roundabout sense, and a year later, with Straus, the weak field equations (WFE) were found:

$$g_{\mu\nu,\lambda} - \Gamma^\sigma_{\mu\lambda}g_{\sigma\nu} - \Gamma^\sigma_{\lambda\nu}g_{\mu\sigma} = 0, \quad \Gamma_\mu = 0,$$
$$R_{(\mu\nu)} = 0, \quad R_{[[\mu\nu],\lambda]} = 0. \tag{5}$$

The last of these implies the existence of a 4-vector of which $R_{[\mu\nu]}$ is a curl but Einstein, by this time, decided that the electromagnetic field tensor should be $g_{[\mu\nu]}$ (the mists of GR !) with (what was worse !) $g_{(\mu\nu)}$ being the old metric tensor (and gravitational potential). Beset by other problems, notable among which was the possibility of representing elementary particles by nonsingular solutions of the field equations, Einstein did not realise the essential uniqueness of the WFE.

Together with Kaufman, he introduced in 1954 new variational parameters instead of the affine connection $\Gamma^{\lambda}{}_{\mu\nu}$ but expressible in their terms, and such that the Ricci tensor became automatically Hermitian with respect to them. Einstein thought that he had a new theory and returned to the Cartan problem of the strength of sets of differential field equations and that is where he left the theory.

Meanwhile, a number of difficulties were shown to be implied by the theory, or rather, by Einstein's physical interpretation of it. I shall give later a detailed discussion of these problems and a demonstration that they are entirely spurious. One of the paradoxical results however must be mentioned at once. Using the Einstein–Infeld–Hoffman (EIH) technique of approximation (see chapter 4), Infeld showed in 1950 that the SFE led to equations of motion of a charged test particle without a Lorentz force. The same result was obtained by Callaway in 1953 for WFE.

My own investigations of GFT had at first little to do with these problems. Russell and I started by attempting to model Ohm's law for which a cylindrically symmetric solution of the field equations was required. A general cylindrical solution of the WFE remains elusive but numerous partial results were found. Of particular interest was the solution (of SFE)

$$g_{44} = g_{11} = g_{22} = -1, \quad g_{33} = -\rho^2(1-(a^2-b^2)\rho^2), \left.\right\}$$
$$g_{23} = -g_{32} = a\rho^2, \quad g_{34} = -g_{43} = b\rho^2, \left.\right\} \tag{6}$$

(written in the system $x^{\mu} = (x^1, x^2, x^3, x^4)$), $\rho = lr + m, a, b, l, m$ constant. In Einstein's scheme this would correspond to a constant electric and magnetic field which did not make sense unless the electromagnetic field tensor was a second derivative of $g_{[\mu\nu]}$.

Accordingly we proposed the identification

$$f_{\mu\nu} = g^{\alpha\beta} g_{[\mu\nu] \; ;\alpha\beta} \cdot \atop {\scriptstyle + - (+)} \tag{7}$$

Actually this $f_{\mu\nu}$ is not skewsymmetric and a better, and equivalent as far as the conclusions go, definition is

$$f_{\mu\nu} = g^{\alpha\beta} g_{[\mu\nu];\alpha\beta}, \atop {\scriptstyle 00 \; (+)} \tag{8}$$

which is skew as commonly required. It turned out that either identification gave gratuitously the solution of the problem of motion by the use of the unmodified EIH technique. The Lorentz force was that proposed *ad hoc* by Treder in 1957. Later I changed my mind about the identification of $f_{\mu\nu}$ as explained in the text, but the result meant that Einstein's theory was again a possibility.

We soon found also that using the most general variational parameters with respect to which the Ricci tensor was Hermitian (and of which the Einstein–Kaufman parameters were a special case), the WFE were the unique set of field equations compatible with the principle of charge conjugation invariance. They now contained the identities

$$\mathfrak{g}^{[\mu\nu]}{}_{,\nu} = 0, \tag{9}$$

and were in fact expressed in terms of Schrödinger's affine connection

$$\tilde{\Gamma}^{\lambda}{}_{\mu\nu} = \Gamma^{\lambda}{}_{\mu\nu} + \tfrac{2}{3}\delta^{\lambda}{}_{\mu}\Gamma_{\nu}, \tag{10}$$

for which $\tilde{\Gamma}_{\mu}$ vanishes identically. Russell proved in general the equivalence of Einstein–Straus and Einstein–Kaufman theories and the problem of the strength of the field equations vanished. The field equations were also shown to be gauge invariant and Γ_{μ} emerged as a candidate for the potential vector, an interpretation which was adopted.

But now a new problem appeared. If $g_{[\mu\nu]}$ was no longer the electromagnetic field there seemed to be no reason why $g_{(\mu\nu)}$ should be the metric. Later, it became clear that it can not be if consistency with charge conjugation–physical basis was to be maintained. Towards the end of 1977, I guessed that a likely way of determining the metric $a_{\mu\nu}$ was to construct Christoffel brackets, equate them to the symmetric part of the affine connection and regard the 'metric hypothesis'

$$\left\{ {\lambda \atop \mu\nu} \right\}_{a} = \tilde{\Gamma}^{\lambda}{}_{(\mu\nu)} \tag{11}$$

as differential equations for $a_{\mu\nu}$. Two things emerged immediately. The first was that a Riemannian space–time provided a sufficient geometrical background for the nonsymmetric theory instead of a new, and very fancy, non-Riemannian geometry. Secondly, the set of possible solutions of the field equations became severely restricted. In fact, for the static, spherically symmetric case only two solutions were left. One could be interpreted as a magnetic octupole (does this mean that according to GFT, magnetic monopoles do not exist?) and the second, the electric solution locally, led unexpectedly to cosmological consequences.

As a result of the contribution of my students, C. Radford and B. T. McInnes, I now know that the hypothesis (11) is the only definition compatible with the principle of Hermitian symmetry.

It is only at this stage that the complete structure of the non-symmetric theory emerged and the weak principle of geometrisation could be formulated and this, in a more logical order, is the subject matter of the present monograph.

2 Field equations of the generalised field theory

1. Principles of GFT: charge conjugation and Hermitian symmetry

The aims of the generalised field theory can be classified into, on the one hand formal or conceptual and on the other, technical. The formal aims are to construct, without necessarily assuming *a priori* that these are gravitation and electromagnetism, a comprehensive account of the physical fields in the macroscopic context of reality and thereby to explain the limitations of general relativity. If the theory should fail, the failure could then be regarded as a proof that these limitations are a fact of Nature and that gravitation alone is to be identified with the geometry of space–time. The latter is a consequence of the principle of equivalence in one or another of its forms. It seems therefore that this principle should be rejected. But equality of inertial and gravitational masses is an exceedingly well-attested empirical fact and if this is all that equivalence means it can not be denied. Hence, rather than rejecting the principle of equivalence, it must be faded away and replaced by something else, something equally meaningful and certain from a physical point of view, as the basis of the theory.

In GR, a Riemannian V_4 is rigidly identified with a world model. It would be possible in constructing a comprehensive field theory to postulate that the model of physics should be some more general geometry, say a nonmetric topological space of this or that kind, a Finsler space, a five-dimensional V_5, or the like. However, without the principle of equivalence, such a procedure would be a stab in the dark. There would be no chance to achieve that intuitive insight which led Einstein to the correct theory of the gravitational field. In particular, the more general the geometry we selected, the more difficult it would be to choose an appropriate form of the field equations. As Einstein remarked, this choice among the set of all possible equations is the central problem of the comprehensive theory.

In the gauge theories there are always the Bianchi identities to serve as the field equations, but the choice of the structure group is arbitrary.

It can not be elevated to the status of a physically meaningful principle. On the other hand, we know two things about the new theory. One is the expected appearance of 'something like' Maxwell's equations and the other that the theory should reduce to GR when Riemannian assumptions are reintroduced. These assumptions are essentially also two, namely the symmetry of the metric tensor $g_{\mu\nu}$ and of the affine connection $\Gamma^\lambda{}_{\mu\nu}$. Together with (the assumption) of vector length invariance under parallel transfer, they ensure that the latter is given by the usual Christoffel brackets $\left\{ \begin{matrix} \lambda \\ \mu\nu \end{matrix} \right\}_g$, (the subscript g meaning that the bracket is constructed from the tensor $g_{\mu\nu}$; it is a necessary notation because later we shall meet the brackets constructed from another nonsingular, symmetric tensor which I shall take as the metric of the background space–time and use $g_{\mu\nu}$, this time nonsymmetric, as a physical field tensor. Confusion in the notation has been one of the bugbears of the nonsymmetric theory.)

Let us consider now what the possible meaning can be of dropping the requirement of symmetry not only of the affine connection $\Gamma^\lambda{}_{\mu\nu}$ but also of the tensor $g_{\mu\nu}$. This is of course what Einstein did without apparently realising that he thereby altered the whole basis of the field theory. The affine connection necessarily refers to geometry, and indeed it is all that we have to start with. If we denote by δh^λ the increment in the components h^λ of a contravariant vector parallel transferred from a point P with coordinates (x^λ) to Q with $(x^\lambda + \delta x^\lambda)$, then

$$\delta h^\lambda \overset{\mathrm{Df}}{=} -\Gamma^\lambda{}_{\mu\nu} h^\mu \delta x^\nu. \tag{1}$$

The formula

$$\delta h_\lambda \overset{\mathrm{Df}}{=} \Gamma^\sigma{}_{\lambda\nu} h_\sigma \delta x^\nu \tag{2}$$

for a covariant vector (with indices raised and lowered with some 'metric' tensor $a_{\mu\nu}$) now implies that the 'length' $(\sqrt{h_\lambda h^\lambda})$ of a vector remains unchanged under parallel transfer. Strictly speaking, we should have written $\delta h^{\overset{\lambda}{+}}$, $\delta h_{\overset{\lambda}{+}}$ in the definitions (1) and (2) respectively, to distinguish them from the equally possible, but different if $\Gamma^\lambda{}_{\mu\nu}$ is nonsymmetric,

$$\delta h^{\overset{\lambda}{-}} = -\Gamma^\lambda{}_{\nu\mu} h^\mu \delta x^\nu, \text{ or } \delta h^{\overset{\lambda}{0}} = -\Gamma^\lambda{}_{(\mu\nu)} h^\mu \delta x^\nu, \tag{3}$$

with similar covariant relations. $\Gamma^\lambda{}_{[\mu\nu]}$ of course is a tensor and combinations such as (1. 5. 3) (i.e. equation (3) of section 5 of chapter 1)

$$\Gamma^{\lambda}{}_{\mu\nu} = \Gamma^{\lambda}{}_{(\mu\nu)} + \Gamma^{\lambda}{}_{[\mu\nu]}$$

have been criticised as mathematically unnatural. However, $\Gamma^{\lambda}{}_{\mu\nu}$ and $\Gamma^{\lambda}{}_{(\mu\nu)}$ are both affine connections and so all the relations in (1) and (3) are meaningful. Integrating expression (1) round a simple, infinitesimal 2-circuit to find the change in a vector traversing it by parallel transfer, we immediately recover the standard definition of the Riemann tensor

$$R^{\lambda}{}_{\mu\nu\sigma} = - \Gamma^{\lambda}{}_{\mu\nu,\sigma} + \Gamma^{\lambda}{}_{\mu\sigma,\nu} + \Gamma^{\rho}{}_{\mu\sigma}\Gamma^{\lambda}{}_{\rho\nu} - \Gamma^{\rho}{}_{\mu\nu}\Gamma^{\lambda}{}_{\rho\sigma}, \tag{4}$$

whether $\Gamma^{\lambda}{}_{\mu\nu}$ is symmetric or not. The Ricci tensor is now obtained by contraction whether or not the geometry is metric:

$$R_{\mu\nu} = R^{\sigma}{}_{\mu\nu\sigma} = - \Gamma^{\sigma}{}_{\mu\nu,\sigma} + \Gamma^{\sigma}{}_{\mu\sigma,\nu} + \Gamma^{\rho}{}_{\mu\sigma}\Gamma^{\sigma}{}_{\rho\nu} - \Gamma^{\rho}{}_{\mu\nu}\Gamma^{\sigma}{}_{\rho\sigma}. \tag{5}$$

Thus, unless we desire a relationship between covariant and contravariant vectors also, geometry itself contains the affine connection and the Riemann and Ricci tensors.

On the other hand, the usual metric relation

$$\mathrm{d}s^2 = g_{\mu\nu}\mathrm{d}x^{\mu}\mathrm{d}x^{\nu} = g_{(\mu\nu)}\mathrm{d}x^{\mu}\mathrm{d}x^{\nu} \tag{6}$$

does not contain the skew part $g_{[\mu\nu]}$ of $g_{\mu\nu}$. If the latter is to be the fundamental tensor of the theory it seems that either the geometry of the physical world must be nonmetric (i.e. purely affine) or $g_{(\mu\nu)}$ must not be the metric tensor. I shall show below that very general mathematical considerations force the second alternative, but in any case, nonmetric physics is difficult to visualise. Any paracompact manifold admits a Riemannian metric and paracompactness is the least we can expect from a physical space–time in which measurement of distance between not-too-widely separated points is possible.

In other words $g_{\mu\nu}$ should not, in the first instance, have anything to do with geometry, which being metric has some other tensor, $a_{\mu\nu}$ say, rather than $g_{(\mu\nu)}$ as the metric tensor. But if correspondence with GR is not to be lost, it is $g_{\mu\nu}$ that is determined by the field equations. Hence, it must represent the physical field. In a four-dimensional world $g_{\mu\nu}$ has sixteen components, and since general relativity tells us that ten functions are needed to describe gravitation, we are left with six functions for the nongravitational fields. Whether this is all the freedom we have depends now on how we determine the affine connection by, presumably, $g_{\mu\nu}$ and its first derivatives. If the relationship is not direct, we may still take the tensor $\Gamma^{\lambda}{}_{[\mu\nu]}$, or at least, the vector Γ_{μ} to be 'physical'. We shall find that again the second case occurs. Thus, physics will be given by $g_{\mu\nu} = g_{(\mu\nu)} + g_{[\mu\nu]}$ and Γ_μ and geometry, by the rest of the connection or rather, by the auxiliary connection (1.5.10)

$$\tilde{\Gamma}^\lambda_{\ \mu\nu} = \Gamma^\lambda_{\ \mu\nu} + \tfrac{2}{3}\delta^\lambda_{\ \mu}\Gamma_\nu.$$

Geometrisation of physics can then be achieved only after the field equations have been found.

I shall refer to this initial bifurcation of physics and geometry and to their final linking in the structure of the total field as the weak principle of geometrisation (WPG). And it now follows that the background Riemannian space will have to be determined after the field equations have been solved by a new relation, perhaps with a partial status of a physical law.

Let us now summarise the principles of GFT. The field equations (FE) are to be chosen so that:

(i) they are independent of any arbitrary motion of the observers or of their choice of a coordinate system (principle of covariance, PC). This is the same as in general relativity but the local transformation group will have to be enlarged if we are not to restrict ourselves unduly.

(ii) They (FE) determine the fundamental structure of the macrophysical laws as well as the geometry of space–time (WPG).

(iii) They (FE) reduce to the general relativistic field equations in the symmetric case (principle of reducibility, PR).

This principle is new and it is not obeyed by GR. A flat Riemannian space may be a Minkowski space but if the field equations (1.4.2) are obeyed, it is empty of all matter and field. On the other hand, PR will be required to formulate the metric hypothesis by which the background V_4 is determined.

The fourth and final principle is the basic hypothesis which gives the nonsymmetric theory its physical meaning and enables the choice of the field equations to be made.

(iv) The field equations are to be Hermitian symmetric with respect to suitably chosen fundamental quantities (principle of Hermitian symmetry or of transposition invariance, PHS).

PHS was interpreted by Einstein as expressing the invariance of the laws of physics under the conjugation of the sign of the electric charge. It foreshadows therefore the emergence of Maxwell's equations as part of the total field we are considering and gives the theory its context. Together with PR, this principle represents the fading away of the principle of equivalence. We must see whether it is adequate.

2. Mathematical formulation of the principle of Hermitian symmetry

It can be observed at once that the way the principle of Hermitian symmetry has been stated above implies that the quantities with respect to which the field equations are to be (Hermitian) symmetric need not be necessarily the field $g_{\mu\nu}$ and the connection $\Gamma^{\lambda}_{\mu\nu}$ which we have so far unambiguously introduced. The criterion is that in some sense both physics and geometry should be represented by the fundamental quantities. Indeed, I shall presently replace the connection by a more suitable pseudo-connection to ensure generality of application of the principle. Restriction of the field representation to a nonsymmetric tensor $g_{\mu\nu}$ is dictated only by a desire to find the simplest extension of general relativity to which the more comprehensive theory should readily reduce through reimposition of ordinary symmetry on the quantities employed.

Nevertheless, the concept of Hermitian symmetry requires some explanation. It has a dual aim: to limit the choice of possible field equations and to model their invariance under charge conjugation. In other words, the idea is that of the well-known PCT theorem of quantum mechanics, the 'C' part is fully macroscopic. Now, clearly, interchange of positive and negative charges is well represented mathematically by complex conjugation. However, if we have

$$a_{\mu\nu} = b_{\mu\nu} + ic_{\mu\nu}, \tag{1}$$

where $b_{\mu\nu}$ and $c_{\mu\nu}$ are real, the Hermitian conjugate of $a_{\mu\nu}$ is

$$*a_{\mu\nu} = b_{\nu\mu} - ic_{\nu\mu}. \tag{2}$$

Hence, the Hermitian symmetry requirement

$$*a_{\mu\nu} = a_{\mu\nu}$$

immediately implies that $b_{\mu\nu}$ is symmetric and $c_{\mu\nu}$, skewsymmetric in the usual sense. But since symmetric and skewsymmetric geometrical objects (that is functions of position on which a transformation of coordinates induces some linear law of transformation) transform independently of each other, we can dispense with the imaginary factor. Thus, although I shall retain the convenient term 'Hermitian', complex quantities will not be used in the sequel. (This is in conformity with Einstein's ideas, except that he referred to the new operation as 'double transposition'; if I continue to call it 'Hermitian symmetry' it is only not to lose its physical relevance as expressing charge conjugation invariance.)

Let us then be given a number of 'fundamental' quantities

$$a, b, c, \ldots, \tag{3}$$

for which the operation of transposition

$$a \to a^{\mathrm{T}}, \text{ etc.} \tag{4}$$

is meaningful. Let a 'transposable' field quantity A be constructed from them:

$$A = A(a, b, c, \ldots). \tag{5}$$

Then the Hermitian conjugate of A is defined by

$$*A = (A(a^{\mathrm{T}}, b^{\mathrm{T}}, c^{\mathrm{T}}, \ldots))^{\mathrm{T}}. \tag{6}$$

A is said to be Hermitian symmetric (or just Hermitian) if

$$*A = A, \tag{7}$$

and skew-Hermitian if

$$*A = -A. \tag{8}$$

Clearly, any such field quantity can be invariably written as a sum of its Hermitian and skew-Hermitian parts:

$$A = \tfrac{1}{2}(*A + A) + \tfrac{1}{2}(A - *A).$$

Since the T-operation (transposition) is really only meaningful for matrices of rank 2, if the fundamental quantities, or for that matter the field quantity A constructed from them, should have more indices than 2, it may be necessary to specify the index pair with respect to which the operation is carried out. This does not create much difficulty in GFT where, at most, expressions like

$$g_{\mu\nu}; \; g_{\mu\nu, \lambda}, \; \Gamma^{\lambda}{}_{\mu\nu, \lambda}, \; \Gamma^{\lambda}{}_{\mu\nu}$$

or products of the latter occur. It is then understood that only covariant indices in the connection or the actual indices of the fundamental g tensor are transposed (the differentiation operator $g_{\mu\nu, \lambda} = \partial_{\lambda} g_{\mu\nu}$ is not regarded of course as fundamental!). With these reservations, it can be easily verified that the double transposition (6) is fully equivalent to the Hermitian conjugation if complex quantities are used.

Let us now consider the Ricci tensor (1.5) (i.e. equation (5) of section 1 of this chapter). Its conjugate with respect to the affine connection is

$$*R_{\mu\nu} = -\Gamma^{\sigma}{}_{\mu\nu, \sigma} + \Gamma^{\sigma}{}_{\sigma\nu, \mu} + \Gamma^{\rho}{}_{\mu\sigma}\Gamma^{\sigma}{}_{\rho\nu} - \Gamma^{\rho}{}_{\mu\nu}\Gamma^{\sigma}{}_{\sigma\rho}, \tag{9}$$

so that the tensor is Hermitian iff

$$\Gamma^{\sigma}{}_{\mu\sigma, \nu} - \Gamma^{\sigma}{}_{\sigma\nu, \mu} - 2\Gamma^{\rho}{}_{\mu\nu}\Gamma_{\rho} = 0, \tag{10}$$

or

$$\Gamma^{\sigma}{}_{(\mu\sigma), \nu} - \Gamma^{\sigma}{}_{(\nu\sigma), \mu} + \Gamma_{\mu; \nu} + \Gamma_{\nu; \mu} = 0. \tag{11}$$

It can be easily checked that this is a tensor equation. If we assume with Einstein that $\Gamma^{\lambda}_{\mu\nu}$ is given in terms of $g_{\mu\nu}$ and its first derivatives by

$$g_{\mu\nu;\lambda} \equiv g_{\mu\nu,\lambda} - \Gamma^{\sigma}_{\mu\lambda}g_{\sigma\nu} - \Gamma^{\sigma}_{\lambda\nu}g_{\mu\sigma} = 0, \tag{12}$$

which is Hermitian symmetric with respect to the index pair μ, ν and with respect to the fundamental (or Hermitian) variables $g_{\mu\nu}$ and $\Gamma^{\lambda}_{\mu\nu}$ (as can be easily checked out), and the tensor $g_{\mu\nu}$ is nonsingular so that there exists a unique tensor $g^{\mu\nu}$ such that

$$g^{\lambda\nu}g_{\mu\nu} = g^{\nu\lambda}g_{\nu\mu} = \delta^{\lambda}_{\mu}, \tag{13}$$

it follows from equation (12) that

$$\Gamma^{\sigma}_{(\mu\sigma)} = \frac{\partial}{\partial x^{\mu}}\ln \sqrt{(-g)}, \tag{14}$$

so that

$$\Gamma^{\sigma}_{(\mu\sigma),\nu} - \Gamma^{\sigma}_{(\nu\sigma),\mu} = 0. \tag{15}$$

These relations are the same as in Riemannian geometry with g replaced by the determinant a of the symmetric metric tensor $a_{\mu\nu}$. With (15), equation (11) reduces to

$$\Gamma_{\mu;\nu} + \Gamma_{\nu;\mu} = 0. \tag{16}$$

Vanishing of Γ_{μ} (which Einstein postulated) ensures, together with the equation (12), that the Ricci tensor is Hermitian symmetric with respect to the field tensor and the affine connection. This is the origin of the strong field equations (1.5.4). But the equations

$$\Gamma_{\mu} = 0 \tag{17}$$

are neither necessary nor physically very clear.

Permuting cyclically the indices μ, ν, λ in equation (12) we easily find that

$$(g^{\varepsilon\sigma}g^{\eta\lambda}g_{\nu\alpha} + g^{\sigma\varepsilon}g^{\lambda\eta}g_{\alpha\nu})\Gamma^{\alpha}_{\sigma\lambda} = g^{\varepsilon\sigma}g^{\eta\lambda}g_{\nu\lambda,\sigma} - g^{\varepsilon\sigma}g^{\lambda\eta}g_{\lambda\sigma,\nu}$$
$$+ g^{\sigma\varepsilon}g^{\lambda\eta}g_{\sigma\nu,\lambda},$$

which on skewsymmetrising with respect to ε and η, and on contracting over η and ν, gives

$$g^{(\mu\nu)}\Gamma_{\mu} = (1/\sqrt{(-g)})(\sqrt{(-g)}g^{[\mu\nu]})_{,\nu}, \tag{18}$$

so that vanishing of Γ_{μ} yields the equations

$$g^{[\mu\nu]}_{,\nu} = 0. \tag{19}$$

Einstein interpreted these equations as the second set of Maxwell equations but his interpretation will be shown to be untenable (see the chapter on motion).

I have mentioned already that the principle of Hermitian symmetry does not require the Hermitian variables to be $g_{\mu\nu}$ and $\Gamma^\lambda{}_{\mu\nu}$ (although it may be difficult to replace the former since it represents our main link with physics). On the other hand, Einstein and Kaufman introduced, instead of the affine connection, the variables $U^\lambda{}_{\mu\nu}$ defined by

$$\Gamma^\lambda{}_{\mu\nu} = U^\lambda{}_{\mu\nu} - \tfrac{1}{3} U^\sigma{}_{\mu\sigma} \delta^\lambda{}_\nu, \quad U^\lambda{}_{\mu\nu} = \Gamma^\lambda{}_{\mu\nu} - \Gamma^\sigma{}_{\mu\sigma} \delta^\lambda{}_\nu. \tag{20}$$

In terms of these, the Ricci tensor is

$$R_{\mu\nu} = - U^\sigma{}_{\mu\nu,\sigma} + U^\sigma{}_{\mu\rho} U^\rho{}_{\sigma\nu} - \tfrac{1}{3} U^\sigma{}_{\mu\sigma} U^\rho{}_{\rho\nu} \tag{21}$$

and is automatically Hermitian with respect to $U^\lambda{}_{\mu\nu}$ which was the motive for their use. However, we can readily convince ourselves that equation (12) is no longer Hermitian, the condition for it to be so becoming

$$\Gamma^\sigma{}_{(\mu\sigma)} = 0, \tag{22}$$

equivalent, by equation (14), to

$$g = \text{constant.} \tag{23}$$

The variables $U^\lambda{}_{\mu\nu}$ are by no means unique in making the Ricci tensor Hermitian and much more general variables of this type will be introduced in the next section. It is interesting however to mention a theorem which throws some light on the uniqueness of the non-symmetric theory. Indeed, there is no linear expression for $\Gamma^\lambda{}_{\mu\nu}$ in terms of new variables $U^\lambda{}_{\mu\nu}$ which satisfies simultaneously the following conditions:

(a)   it is solvable for $U^\lambda{}_{\mu\nu}$ in terms of $\Gamma^\lambda{}_{\mu\nu}$;

(b)   the condition that $g_{\mu\nu;\lambda}$ should be Hermitian with respect to $g_{\mu\nu}$
      $\underset{+\ -}{}$
      and $U^\lambda{}_{\mu\nu}$ is a tensor (or invariant) equation;

(c) $- \Gamma^\sigma{}_{\mu\nu,\sigma} + \Gamma^\sigma{}_{\mu\sigma,\nu}$ (i.e. the derivative dependant part of the Ricci tensor) is Hermitian in $U^\lambda{}_{\mu\nu}$.

The proof of this theorem is elementary but somewhat tedious and will not be given here. The result does not mean that Hermitian variables other than the affine connection must not be used. Surprisingly, it is the condition (a) that lets us down. What we shall find is that even the most general Hermitian variables, which guarantee the consistency of the theory with PHS, lead to a unique set of field equations when these are expressed in terms of a new affine connection which is only one-sidedly related to the connection $\Gamma^\lambda{}_{\mu\nu}$ (the algebraic expression can not be inverted). The equations (19) are identities of GFT.

3.   Variational parameters and the field equations

Relativity, and *a fortiori* GFT which we are discussing, differ from most other branches of theoretical physics in that one gropes within them for an understanding of the fundamental structure of the world (though not for the ultimate structure of matter which is the province of quantum mechanics), rather than for an understanding of empirical evidence or an ordering of observational data. GFT is therefore closely related to thought processes, to logic and to philosophy although its language is mathematical. A comparison with empirical knowledge comes later, after the theory has been constructed.

Now from a logical point of view it makes little difference whether the field equations of a theory are postulated or derived from a variational principle. The latter also must be postulated since there is no known method for calculating a Lagrangian for an essentially unknown theory. On the other hand, it is generally easier to lay down a credible variational principle, or to amend a pre-existing one in a modification of a theory, than to play with complete sets of field equations. In particular, variational methods allow the group symmetries to be surveyed and ensure that spurious variables are not introduced. They also lead automatically to the identities and conservation laws of the theory, though the latter may sometimes have dubious physical meaning. It was Gustav Mie, before the advent of general relativity, who first emphasised the desirability of basing a fundamental theory on a variational principle. We shall find when we come to the problem of defining the Riemannian metric of GFT that very general, mathematical considerations do not leave as much freedom in the choice of the Lagrangian as might have been expected.

Nevertheless, there is a freedom which we always have and that is in choosing the parameters with respect to which the variation is to be carried out. Freedom of choice here should be interpreted as preservation of as much generality as is consistent with the principles of the theory. I have already stated that it is difficult to avoid using $g_{\mu\nu}$ (or $g^{\mu\nu}$, or $\mathfrak{g}^{\mu\nu}$). Instead of $\Gamma^\lambda_{\mu\nu}$, however, we shall select variables $W^\lambda_{\mu\nu}$, say, which satisfy the following conditions:

(*a*)   the relation $W^\lambda_{\mu\nu} = W^\lambda_{\mu\nu}(\Gamma^\alpha_{\beta\gamma})$ is invertible;

and

(*b*) $R_{\mu\nu}(W^\alpha_{\beta\gamma})$ is identically Hermitian with respect to $W^\lambda_{\mu\nu}$.

It can be shown easily that the most general linear relationship between the new variables and the components of the affine connection is then

$$\Gamma^\lambda{}_{\mu\nu} = W^\lambda{}_{\mu\nu} + \tfrac{1}{5}(2\lambda_1 - 1)\delta^\lambda{}_\nu W^\sigma{}_{(\mu\sigma)} - \tfrac{1}{3}\delta^\lambda{}_\nu W_\mu$$
$$- \tfrac{1}{5}(3\lambda_1 + 1)W^\sigma{}_{(\nu\sigma)}\delta^\lambda{}_\mu + \tfrac{1}{3}(\lambda_2 + 1)\delta^\lambda{}_\mu W_\nu, \tag{1}$$

where

$$W_\mu = W^\sigma{}_{[\mu\sigma]}$$

and

$$W^\lambda{}_{\mu\nu} = \Gamma^\lambda{}_{\mu\nu} - \frac{1}{30\lambda_1\lambda_2}[(6\lambda_1\lambda_2 - 15\lambda_1 - 3\lambda_2)\delta^\lambda{}_\nu\Gamma^\sigma{}_{(\mu\sigma)}$$
$$+ (10\lambda_1\lambda_2 - 5\lambda_1 - 5\lambda_2)\delta^\lambda{}_\nu\Gamma_\mu$$
$$+ (6\lambda_1\lambda_2 + 15\lambda_1 - 3\lambda_2)\delta^\lambda{}_\mu\Gamma^\sigma{}_{(\nu\sigma)}$$
$$- (10\lambda_1\lambda_2 - 5\lambda_1 + 5\lambda_2)\delta^\lambda{}_\mu\Gamma_\nu], \tag{2}$$

where λ_1, λ_2 are constants subject only to the condition

$$\lambda_1\lambda_2 \neq 0. \tag{3}$$

In terms of $W^\lambda{}_{\mu\nu}$, the Ricci tensor becomes

$$R_{\mu\nu} = - W^\sigma{}_{\mu\nu,\sigma} + \tfrac{1}{30}(9\lambda_1 + 5\lambda_2 + 8)(W^\sigma{}_{\mu\sigma,\nu} + W^\sigma{}_{\sigma\nu,\mu})$$
$$- \tfrac{1}{30}(5\lambda_2 - 9\lambda_1 + 2)(W^\sigma{}_{\sigma\mu,\nu} + W^\sigma{}_{\nu\sigma,\mu}) + W^\sigma{}_{\mu\rho}W^\rho{}_{\sigma\nu}$$
$$- \tfrac{2}{5}(3\lambda_1 + 1)W^\sigma{}_{\mu\nu}W^\rho{}_{(\sigma\rho)} - \tfrac{1}{75}(3\lambda_1 - 4)^2 W^\sigma{}_{\mu\sigma}W^\rho{}_{\rho\nu}$$
$$- \tfrac{1}{75}(3\lambda_1 - 4)(3\lambda_1 + 1)(W^\sigma{}_{\mu\sigma}W^\rho{}_{\nu\rho} + W^\sigma{}_{\sigma\mu}W^\rho{}_{\rho\nu}) \tag{4}$$
$$- \tfrac{1}{75}(3\lambda_1 + 1)^2 W^\sigma{}_{\sigma\mu}W^\rho{}_{\nu\rho}.$$

We may note that the Einstein–Kaufman U-variables are recovered if

$$\lambda_1 = -\tfrac{1}{3}, \quad \lambda_2 = -1. \tag{5}$$

Let us also assume that the field equations are to be derived from the Hermitian variational principle

$$\delta \int \mathfrak{g}^{\mu\nu} R_{\mu\nu}(W^\alpha{}_{\beta\gamma}) = 0 \tag{6}$$

under the standard assumption that all integrated parts (3-integrals over hypersurfaces bounding the four-dimensional volume of integration) vanish on the boundary. Variation in $\mathfrak{g}^{\mu\nu}$ gives immediately the sixteen equations (differential field equations)

$$R_{\mu\nu}(W^\alpha{}_{\beta\gamma}) = 0. \tag{7}$$

Substituting into these from equation (2), we get (independently of λ_1, and λ_2)

$$- \Gamma^\sigma{}_{(\mu\nu),\sigma} + \tfrac{1}{2}(\Gamma^\sigma{}_{(\mu\sigma),\nu} + \Gamma^\sigma{}_{(\nu\sigma),\mu}) + \Gamma^\rho{}_{(\mu\sigma)}\Gamma^\sigma{}_{(\rho\nu)} - \Gamma^\rho{}_{(\mu\nu)}\Gamma^\sigma{}_{(\rho\sigma)}$$
$$= - \tfrac{1}{2}(\Gamma_{\mu,\nu} + \Gamma_{\nu,\mu}) - \Gamma^\rho{}_{[\mu\sigma]}\Gamma^\sigma{}_{[\rho\nu]} + \Gamma^\rho{}_{[\mu\nu]}\Gamma_\rho \tag{8}$$

and

$$-\Gamma^{\sigma}_{[\mu\nu],\sigma} + \tfrac{1}{2}(\Gamma_{\mu,\nu} - \Gamma_{\nu,\mu}) + \Gamma^{\rho}_{[\mu\sigma]}\Gamma^{\sigma}_{[\rho\nu]} - \Gamma^{\rho}_{[\mu\nu]}\Gamma^{\sigma}_{\rho\sigma}$$
$$= \tfrac{1}{2}(\Gamma^{\sigma}_{(\nu\sigma),\mu} - \Gamma^{\sigma}_{(\mu\sigma),\nu}). \tag{9}$$

Similarly, if we independently vary $W^{\lambda}_{\mu\nu}$, we get the sixty-four equations (algebraic equations determining $W^{\lambda}_{\mu\nu}$ and therefore also the affine connection in terms of $g_{\mu\nu}$ and its first derivatives)

$$g^{\mu\nu}{}_{,\lambda} - (\mu_1 + \mu_2)(g^{\mu\sigma}{}_{,\sigma}\delta^{\nu}{}_{\lambda} + g^{\sigma\nu}{}_{,\sigma}\delta^{\mu}{}_{\lambda}) + \mu_2(g^{\nu\sigma}{}_{,\sigma}\delta^{\mu}{}_{\lambda}$$
$$+ g^{\sigma\mu}{}_{,\sigma}\delta^{\nu}{}_{\lambda}) + g^{\mu\sigma}W^{\nu}{}_{\lambda\sigma} + g^{\sigma\nu}W^{\mu}{}_{\sigma\lambda} - 2\mu_1 g^{\mu\nu}W^{\sigma}{}_{(\lambda\sigma)}$$
$$- \mu_1 g^{\rho\sigma}(W^{\mu}{}_{\rho\sigma}\delta^{\nu}{}_{\lambda} + W^{\nu}{}_{\rho\sigma}\delta^{\mu}{}_{\lambda}) - \tfrac{1}{3}(\mu_1 - 1)^2(g^{\mu\sigma}W^{\rho}{}_{\rho\sigma}\delta^{\nu}{}_{\lambda}$$
$$+ g^{\sigma\nu}W^{\rho}{}_{\sigma\rho}\delta^{\mu}{}_{\lambda}) - \tfrac{1}{3}\mu^2{}_1(g^{\nu\sigma}W^{\rho}{}_{\sigma\rho}\delta^{\mu}{}_{\lambda} + g^{\sigma\mu}W^{\rho}{}_{\rho\sigma}\delta^{\nu}{}_{\lambda}) \tag{10}$$
$$- \tfrac{1}{3}\mu_1(\mu_1 - 1)(g^{\mu\sigma}W^{\rho}{}_{\rho\sigma}\delta^{\nu}{}_{\lambda} + g^{\sigma\mu}W^{\rho}{}_{\sigma\rho}\delta^{\nu}{}_{\lambda} + g^{\sigma\nu}W^{\rho}{}_{\rho\sigma}\delta^{\mu}{}_{\lambda}$$
$$+ g^{\nu\sigma}W^{\rho}{}_{\sigma\rho}\delta^{\mu}{}_{\lambda}) = 0,$$

where we have put

$$5\mu_1 = 3\lambda_1 + 1 \text{ and } 6\mu_2 = 1 + \lambda_2 - 3\mu_1, \tag{11}$$

(so that the Einstein–Kaufman case is $\mu_1 = \mu_2 = 0$). Skewsymmetrising equations (10) in μ and ν, and contracting over ν and λ, we get

$$-(3\mu_1 - 1 + 6\mu_2)g^{[\mu\nu]}{}_{,\nu} = -\lambda_2 g^{[\mu\nu]}{}_{,\nu} = 0, \tag{12}$$

whence, because of the condition (3), we obtain the expected identities

$$g^{[\mu\nu]}{}_{,\nu} = 0. \tag{13}$$

Let us now rewrite equations (10) in terms of the affine connection. One point should be mentioned. I have already tacitly assumed that

$$\det(g_{\mu\nu}) < 0. \tag{14}$$

Although I do not, emphatically, regard $g_{(\mu\nu)}$ as the metric tensor, the principle of reducibility demands this assumption. Contracting now equations (10) over ν and λ, and substituting the result back into themselves we get, because of the identities (13),

$$g^{\mu\nu}{}_{,\lambda} + g^{\mu\sigma}W^{\nu}{}_{\lambda\sigma} + g^{\sigma\nu}W^{\mu}{}_{\sigma\lambda} - 2\mu_1 g^{\mu\nu}W^{\sigma}{}_{(\lambda\sigma)}$$
$$+ \tfrac{1}{3}[(2\mu_1 - 1)g^{\mu\sigma}W^{\rho}{}_{(\sigma\rho)} + g^{\mu\sigma}W_{\sigma}]\delta^{\nu}{}_{\lambda}$$
$$+ \tfrac{1}{3}[(2\mu_1 - 1)g^{\sigma\nu}W^{\rho}{}_{(\sigma\rho)} - g^{\sigma\nu}W_{\sigma}]\delta^{\mu}{}_{\lambda} = 0.$$

Using (2.13) (implied, of course, by the condition (13) above) we can easily lower the indices, and cancelling out $\sqrt{(-g)}$ and substituting from the equation (2), we finally find that

$$g_{\mu\nu,\lambda} - \Gamma^{\sigma}{}_{\mu\lambda}g_{\sigma\nu} - \Gamma^{\sigma}{}_{\lambda\nu}g_{\mu\sigma} - \tfrac{2}{3}(\Gamma_{\lambda}g_{\mu\nu} + \Gamma_{\nu}g_{\mu\lambda}) = 0.$$

This can be further rewritten as

$$g_{\mu\nu,\lambda} - (\Gamma^\sigma_{\mu\lambda} + \tfrac{2}{3}\delta^\sigma_\mu \Gamma_\lambda)g_{\sigma\nu} - (\Gamma^\sigma_{\lambda\nu} + \tfrac{2}{3}\delta^\sigma_\lambda \Gamma_\nu)g_{\mu\sigma} = 0. \tag{15}$$

Let us now introduce the auxiliary (Schrödinger) connection

$$\tilde{\Gamma}^\lambda_{\mu\nu} = \Gamma^\lambda_{\mu\nu} + \tfrac{2}{3}\delta^\lambda_\mu \Gamma_\nu. \tag{16}$$

In terms of it, the Ricci (Schrödinger) tensor

$$\begin{aligned}
R_{\mu\nu}(\tilde{\Gamma}) &= -\tilde{\Gamma}^\sigma_{\mu\nu,\sigma} + \tilde{\Gamma}^\sigma_{\mu\sigma,\nu} + \tilde{\Gamma}^\sigma_{\mu\rho}\tilde{\Gamma}^\rho_{\sigma\nu} - \tilde{\Gamma}^\sigma_{\mu\nu}\tilde{\Gamma}^\rho_{\sigma\rho} \\
&= R_{\mu\nu}(\Gamma) - \tfrac{2}{3}(\Gamma_{\nu,\mu} - \Gamma_{\mu,\nu}),
\end{aligned}$$

where $R_{\mu\nu}(\Gamma)$ is the Ricci tensor (4) expressed in terms of the connection $\Gamma^\lambda_{\mu\nu}$. The field equations (15) now become

$$\left.\begin{aligned}
& g_{\mu\nu;\lambda}(\tilde{\Gamma}) = 0, \quad \tilde{\Gamma}_\mu = 0, \\
& {}_{+-} \\
& R_{(\mu\nu)}(\tilde{\Gamma}) = 0,
\end{aligned}\right\} \tag{17a}$$

and

$$R_{[\mu\nu]}(\tilde{\Gamma}) = -\tfrac{2}{3}(\Gamma_{\nu,\mu} - \Gamma_{\mu,\nu}) \tag{17b}$$

or, eliminating Γ_μ,

$$R_{[[\mu\nu],\lambda]}(\tilde{\Gamma}) = 0. \tag{17c}$$

These are the weak field equations of Einstein and Straus which we have now shown to be the most general set of field equations consistent with the principle of Hermitian symmetry and with the variational principle (6).

4. Why weak field equations?

The equations (3.17) must now be assumed to determine the field laws of macrophysics as well as the geometry although we do not yet know how. Because they are independent of the parameters λ_1, λ_2 or $W^\lambda_{\mu\nu}$, these variables appear to be purely auxilliary with no direct, physical or geometrical meaning other than expressing general conformity to PHS. Nevertheless, we shall find that this is not quite so (see section 6). At least λ_2 does have a bearing on the physical meaning of the theory. We may observe that the field equations constitute apparently $64 + 4 + 16 = 84$ equations for only 16 $(g_{\mu\nu}) + 64$ $(\Gamma^\lambda_{\mu\nu})$ $= 80$ unknowns but because of their derivation from a variational principle, we are assured of their formal compatability.

Their name 'weak field equations' arose out of the early, and erroneous, identification of $g_{\mu\nu}$ with the gravitational and of $g_{[\mu\nu]}$ with the electromagnetic field. Einstein argued that in that case one could consider a weak field approximation

$$g_{\mu\nu} = \eta_{\mu\nu} + h_{\mu\nu} + k_{\mu\nu}, \tag{1}$$

where

$$h_{\mu\nu} = h_{\nu\mu}, \quad k_{\mu\nu} = -k_{\nu\mu} \tag{2}$$

are small in the sense that $h^2 \ll h$, $k^2 \ll k$. Actually, the requirement (1) has less to do with the physical interpretation of the field equations than with the principle of reducibility. If we assume the approximation (1) this splits the symmetric and skewsymmetric parts of the equations. Thus, if

$$\tilde{\Gamma}^{\lambda}{}_{\mu\nu} = \underset{2}{\tilde{\Gamma}}{}^{\lambda}{}_{\mu\nu} + \underset{1}{\tilde{\Gamma}}{}^{\lambda}{}_{\mu\nu} + \underset{2}{\tilde{\Gamma}}{}^{\lambda}{}_{\mu\nu} + \dots, \tag{3}$$

where 0 and 1 denote orders of magnitude, we readily find that

$$\underset{0}{\tilde{\Gamma}}{}^{\lambda}{}_{\mu\nu} = 0, \tag{4}$$

and

$$\underset{1}{\tilde{\Gamma}}{}^{\lambda}{}_{(\mu\nu)} = \tfrac{1}{2}\eta^{\lambda\sigma}(h_{\sigma\nu,\mu} + h_{\mu\sigma,\nu} - h_{\mu\nu,\sigma}), \quad \underset{1}{\tilde{\Gamma}}{}^{\lambda}{}_{[\mu\nu]} = \tfrac{1}{2}\eta^{\lambda\sigma}k_{[\sigma\nu,\mu]}. \tag{5}$$

Because of (4)

$$\underset{1}{R}_{\mu\nu}(\tilde{\Gamma}) = -\underset{1}{\tilde{\Gamma}}{}^{\sigma}{}_{\mu\nu,\sigma} + \underset{1}{\tilde{\Gamma}}{}^{\sigma}{}_{\mu\sigma,\nu},$$

and

$$\underset{1}{R}_{[\mu\nu]}(\tilde{\Gamma}) = -\underset{1}{\tilde{\Gamma}}{}^{\sigma}{}_{[\mu\nu],\sigma} = -\tfrac{1}{2}\eta^{\lambda\sigma}k_{[\sigma\nu,\mu],\lambda}. \tag{6}$$

If we now consider the last of the field equations (17) (in the form independent of Γ_μ), we see that it becomes

$$\tfrac{1}{2}\eta^{\rho\sigma}(k_{[\mu\nu,\lambda]})_{,\rho\sigma} = \square\, k_{[\mu\nu,\lambda]} = 0, \tag{7}$$

where the operator $\square$ denotes as usual the D'Alembertian. These are 'weaker' than Maxwell's equations which presumably would be

$$k_{\mu\nu,\lambda} = 0.$$

Hence WFE. But they are also weaker than SFE given by (1.5.4). The two sets coincide if

$$\Gamma_\mu = 0,$$

when of course, the geometrical connection becomes identical with Schrödinger's connection. We shall see later that the SFE can not lead to equations of motion of a charged test particle with a Lorentz force. From then on we shall be able to abandon any consideration of this set once and for all as incapable of describing fundamental macrophysics.

5. Uniqueness and gauge invariance of WFE

Einstein and Straus derived the weak field equations from the variational principle

$$\delta \int (\mathfrak{g}^{\mu\nu} \bar{R}_{\mu\nu} + 2\alpha_\mu \mathfrak{g}^{[\mu\nu]}{}_{,\nu}) = 0. \tag{1}$$

Here α_μs are the Lagrange multipliers playing no part in the variation and which had to be eliminated from the final field equations.

$$\bar{R}_{\mu\nu} = - \Gamma^\sigma{}_{\mu\nu,\sigma} + \tfrac{1}{2}(\Gamma^\sigma{}_{\mu\sigma,\nu} + \Gamma^\sigma{}_{\sigma\nu,\mu}) + \Gamma^\sigma{}_{\mu\rho}\Gamma^\rho{}_{\sigma\nu} - \Gamma^\sigma{}_{\mu\nu}\Gamma^\rho{}_{(\sigma\rho)}$$

is that part of the Ricci tensor which is Hermitian with respect to the affine connection.

Variation of (1) with respect to $\mathfrak{g}^{\mu\nu}$ gives

$$\bar{R}_{\mu\nu} = \alpha_{\mu,\nu} - \alpha_{\nu,\mu},$$

so that

$$\bar{R}_{(\mu\nu)} = 0, \quad \bar{R}_{[[\mu\nu],\lambda]} = 0.$$

Also, variation in $\Gamma^\lambda{}_{\mu\nu}$ under the condition (which is consistent here),

$$\Gamma_\mu = 0,$$

gives

$$g_{\mu\nu;\lambda}(\Gamma) = 0.$$
$$\phantom{g_{\mu\nu}}_{+\,-}$$

Complete equivalence of WFE with the Einstein–Kaufman theory, and therefore with GFT, was proved by G. K. Russell. Let us write, in fact the WFE again as

$$g_{\mu\nu;\lambda}(\tilde{\Gamma}) = 0, \quad R_{(\mu\nu)}(\tilde{\Gamma}) = 0, \quad R_{[\mu\nu]}(\tilde{\Gamma}) = \alpha_{\mu,\nu} - \alpha_{\nu,\mu}, \quad \tilde{\Gamma}_\mu = 0 \tag{2}$$
$$\phantom{g_{\mu\nu}}_{+\,-}$$

where α_μ is an arbitrary vector. Let also

$$\tilde{\Gamma}^\lambda{}_{\mu\nu} = \gamma^\lambda{}_{\mu\nu} + \alpha_\nu \delta^\lambda{}_\mu.$$

Then $\gamma^\lambda{}_{\mu\nu}$ is still an affine connection. In terms of it the field equations (2) become

$$g_{\mu\nu;\lambda}(\gamma) = \alpha_\lambda g_{\mu\nu} + \alpha_\nu g_{\mu\lambda},$$
$$\phantom{g_{\mu\nu}}_{+\,-}$$

$$R_{(\mu\nu)}(\gamma) = 0 = R_{[\mu\nu]}(\gamma), \quad \gamma_\mu - \tfrac{3}{2}\alpha_\mu = 0.$$

The last of these now gives again the WFE (3.17). We can call the above a hybrid theory. But γ_μ has nothing to do with the theory of Einstein and Straus: we are only asking that the vector α_μ should be the contracted skew part of some affine connection. We obtain GFT if we select $\gamma^\lambda{}_{\mu\nu}$ to be the affine connection of that theory.

Let us consider next whether GFT allows gauge invariance of the type of which much is made in contemporary physics. I have mentioned in section 3 that the Einstein–Kaufman 'theory' results if

$$\lambda_1 = -\tfrac{1}{3} \text{ but } \lambda_2 = 1.$$

Denoting the corresponding (Einstein–Kaufman) Hermitian parameters by $U^\lambda{}_{\mu\nu}$ as before, we have

$$U^\lambda{}_{\mu\nu} = W^\lambda{}_{\mu\nu} - \tfrac{1}{3}(3\lambda_1 + 1)(\delta^\lambda{}_\nu W^\sigma{}_{(\mu\sigma)} + \delta^\lambda{}_\mu W^\sigma{}_{(\nu\sigma)})$$
$$+ \tfrac{1}{3}(\lambda_2 + 1)(\delta^\lambda{}_\mu W_\nu - \delta^\lambda{}_\nu W_\mu). \tag{3}$$

If therefore we choose

$$\lambda_1 = -\tfrac{1}{3} \text{ but } \lambda_2 \neq -1, 0,$$

and if there exists a scalar w, such that

$$W_\mu = [w/(\lambda_2 + 1)]_{,\mu}, \tag{4}$$

the field equations are readily seen to be invariant under the substitution

$$U^\lambda{}_{\mu\nu} = W^\lambda{}_{\mu\nu} + \tfrac{1}{3}(\delta^\lambda{}_\mu w_\nu - \delta^\lambda{}_\nu w_\mu), \tag{5}$$

where

$$w_\mu = w_{,\mu}.$$

This is the required form of gauge invariance in GFT.

6. The identities and conservation laws of GFT

Let us now return to the variational principle (3.6)

$$\delta \int \mathfrak{g}^{\mu\nu} R_{\mu\nu}(W^\alpha{}_{\beta\gamma}) = 0,$$

with the Ricci tensor given in terms of $W^\lambda{}_{\mu\nu}$ and its derivatives by the expression (3.4). Although we have seen that the theories obtained by choosing different values for the parameters λ_1 and λ_2 are formally equivalent, an infinitesimal variation in these parameters may lead to a meaningful conservation law.

The equation (3.15) expressed in terms of the auxilliary variables $W^\lambda{}_{\mu\nu}$ is

$$g_{\mu\nu,\lambda} - W^\sigma{}_{\mu\lambda} g_{\sigma\nu} - W^\sigma{}_{\lambda\nu} g_{\mu\sigma} - \tfrac{1}{3}(2\lambda_1 - 1)(2g_{\mu\nu} W^\sigma{}_{(\lambda\sigma)} + g_{\mu\lambda} W^\sigma{}_{(\nu\sigma)}$$
$$+ g_{\lambda\nu} W^\sigma{}_{(\mu\sigma)}) - \tfrac{1}{3}(g_{\mu\lambda} W_\nu - g_{\lambda\nu} W_\mu) = 0. \tag{1}$$

It follows that if a solution for $W^\lambda{}_{\mu\nu}$ exists, it will depend on $g_{\mu\nu}, g_{\mu\nu,\lambda}$ and λ_1, but not on λ_2. Hence variation in λ_1 is not generally independent of the variation in $W^\lambda{}_{\mu\nu}$. On the other hand, we can vary λ_2 with impunity.

A glance at (3.4) now shows that

$$\delta_{\lambda_2} \int g^{\mu\nu} R_{\mu\nu} = \tfrac{1}{3} \int g^{\mu\nu}(W_{\mu,\nu} - W_{\nu,\mu})\delta\lambda_2, \tag{2}$$

so that we obtain the first conservation law

$$g^{[\mu\nu]}W_{\mu,\nu} = 0, \tag{3}$$

which, because of the identities (3.13) can be written as

$$(g^{[\mu\nu]}W_{\mu})_{,\nu} = 0. \tag{4}$$

From the equation (3.2) we immediately find that

$$2\lambda_2 W_\mu = 3\Gamma^\sigma_{(\mu\sigma)} + \Gamma_\mu, \tag{5}$$

while contraction of (3.15) with $g^{\mu\nu}$ gives

$$\Gamma^\sigma_{(\mu\sigma)} + \tfrac{5}{3}\Gamma_\mu = (\ln \sqrt{(-g)})_{,\mu}. \tag{6}$$

Since

$$\lambda_2 \neq 0,$$

by hypothesis, and from (6), $\Gamma^\sigma_{(\mu\sigma),\nu} - \Gamma^\sigma_{(\nu\sigma),\mu}$ is a tensor, the conservation law (4) is an invariant equation (as is well known, infinitesimal methods do not always yield invariant results). Indeed, it reduces to

$$(g^{[\mu\nu]}\Gamma_\mu)_{,\nu} = 0, \tag{7}$$

or to

$$g^{[\mu\nu]}f_{\mu\nu} = 0, \tag{8}$$

where the tensor $f_{\mu\nu}$ is proportional to the curl of Γ_μ. Full physical interpretation of GFT can be discussed only after we have found solutions of the field equations (chapter 5) and, this being peculiar to GFT, solved the problem of motion (chapter 4).

However, $g^{[\mu\nu]}\Gamma_\mu$ being a vector density, it is already tempting to identify it with the density of an electric current

$$\mathfrak{J}^\nu = g^{[\mu\nu]}\Gamma_\mu. \tag{9}$$

If on the other hand, $f_{\mu\nu}$ were to be regarded (as we shall eventually assume) as the electromagnetic field tensor, then Γ_μ would have to be proportional to the 4-vector potential, and both the equations (8) and (9) would indicate that we are dealing with an electromagnetic theory of probably a nonlinear type rather than with a pure Maxwell field. These conclusions can only be avoided if λ_2 is a universal constant but in that case its precise meaning is very difficult to determine.

Following the methods of Weyl and Einstein we can derive other conservation relations of GFT. In particular, the transformation law of an affine connection shows that the coordinate transformation

$$x'^\lambda = x'^\lambda(x^\mu) \left(\det\left(\frac{\partial x'^\lambda}{\partial x^\mu} \right) \neq 0 \right)$$

induces for $W^\lambda{}_{\mu\nu}$, the law

$$W'^\lambda{}_{\mu\nu} = \frac{\partial x'^\lambda}{\partial x^\alpha} \frac{\partial x^\beta}{\partial x'^\mu} \frac{\partial x^\gamma}{\partial x'^\nu} W^\alpha{}_{\beta\gamma} + \frac{\partial x'^\lambda}{\partial x^\alpha} \frac{\partial^2 x^\alpha}{\partial x'^\mu \partial x'^\nu}$$
$$+ \frac{1}{\lambda_1 \lambda_2} \left(P\delta^\lambda{}_\nu \frac{\partial x'^\sigma}{\partial x^\alpha} \frac{\partial^2 x^\alpha}{\partial x'^\sigma \partial x'^\mu} + Q\delta^\lambda{}_\mu \frac{\partial x'^\sigma}{\partial x^\alpha} \frac{\partial^2 x^\alpha}{\partial x'^\sigma \partial x'^\nu} \right), \tag{10}$$

where

$$Q = \tfrac{1}{15}(\lambda_2 - 5\lambda_1 - 2\lambda_1\lambda_2) = P - \lambda_1 - \tfrac{2}{3}. \tag{11}$$

The expression (10) shows that only if either

$$\lambda_1 = -\tfrac{2}{3} \text{ or } \lambda_1 = \lambda_2/(5 + 2\lambda_2)$$

is $W^\lambda{}_{[\mu\nu]}$ a tensor and W_μ a vector.

If we consider now an infinitesimal coordinate transformation

$$x'^\lambda = x^\lambda + \varepsilon\xi^\lambda(x), \quad \varepsilon^2 \ll \varepsilon, \tag{12}$$

the transformation law (10) gives

$$\delta W^\lambda{}_{\mu\nu} = W^\sigma{}_{\mu\nu}\xi^\lambda{}_{,\sigma} - W^\lambda{}_{\sigma\nu}\xi^\sigma{}_{,\mu} - W^\lambda{}_{\mu\sigma}\xi^\sigma{}_{,\nu} - \xi^\lambda{}_{,\mu\nu}$$
$$- \frac{1}{\lambda_1\lambda_2}(P\delta^\lambda{}_\nu\xi^\sigma{}_{,\mu\sigma} + Q\delta^\lambda{}_\mu\xi^\sigma{}_{,\nu\sigma}) - W^\lambda{}_{\mu\nu,\sigma}\xi^\sigma, \tag{13}$$

and

$$\delta g^{\mu\nu} = g^{\sigma\nu}\xi^\mu{}_{,\sigma} + g^{\mu\sigma}\xi^\nu{}_{,\sigma} - g^{\mu\nu}\xi^\sigma{}_{,\sigma} - g^{\mu\nu}{}_{,\sigma}\xi^\sigma \tag{14}$$

to the first order in ε. The last terms in (13) and (14) respectively are added of course, because the transformation (12) can be viewed as a mapping of the manifold onto itself, a neighbourhood being mapped onto an adjacent neighbourhood.

Substituting the expressions (13) and (14) into the variation

$$\int (R_{\mu\nu}\delta g^{\mu\nu} + \mathcal{N}^{\mu\nu}{}_\lambda \delta W^\lambda{}_{\mu\nu}) = 0 \tag{15}$$

in which the expression for $\mathcal{N}^{\mu\nu}{}_\lambda$ analogous to (3.10) can be easily written down using (3.4) (I omit it because it is not relevant to our discussion), we obtain for an arbitrary ξ^λ

$$[R_{\lambda\mu}g^{\nu\mu} + R_{\mu\lambda}g^{\mu\nu} + \mathcal{N}^{\mu\sigma}{}_\lambda W^\nu{}_{\mu\sigma} - \mathcal{N}^{\nu\mu}{}_\sigma W^\sigma{}_{\lambda\mu} - \mathcal{N}^{\mu\nu}{}_\sigma W^\sigma{}_{\mu\lambda}$$
$$+ \mathcal{N}^{\mu\nu}{}_{\lambda,\mu} + \frac{1}{\lambda_1\lambda_2}(P\mathcal{N}^{\nu\sigma}{}_{\sigma,\lambda} + Q\mathcal{N}^{\sigma\nu}{}_{\sigma,\lambda})]_{,\nu} - R_{\mu\nu,\lambda}g^{\mu\nu}$$
$$+ \mathcal{N}^{\mu\nu}{}_\sigma W^\sigma{}_{\mu\nu,\lambda} = 0. \tag{16}$$

These equations are analogous to the contracted Bianchi identities of GR. They ensure freedom of choice of the coordinate system and the general compatibility of the field equations. In GR, of course, the corresponding equations are

$$G^{\mu\nu}_{\ ;\nu} = 0, \tag{17}$$

$G^{\mu\nu}$ being the contravariant Einstein tensor.

A pseudo-conservation law can also be derived as follows. Let

$$\mathscr{L} = \mathfrak{g}^{\mu\nu} R_{\mu\nu}(W). \tag{18}$$

Then, without integrating by parts as was required above to get rid of the derivatives of ξ^λ, we have

$$\delta\mathscr{L} = (-\mathfrak{g}^{\mu\nu}\delta W^\sigma_{\ \mu\nu} + \tfrac{1}{5}(3\lambda_1 + 1)(\mathfrak{g}^{\mu\sigma}\delta W^\nu_{\ \mu\nu} + \mathfrak{g}^{\sigma\mu}\delta W^\nu_{\ \nu\mu})$$
$$+ \tfrac{2}{15}(5\lambda_2 - 9\lambda_1 + 2)\mathfrak{g}^{[\mu\sigma]}\delta W_\mu)_{,\sigma}. \tag{19}$$

If now ξ^λ is assumed to be independent of the coordinates and

$$\delta\mathscr{L} = 0, \tag{20}$$

we get

$$\mathfrak{T}^\mu_{\ \nu,\mu} = 0 \tag{21}$$

where

$$\mathfrak{T}^\mu_{\ \nu} = \mathfrak{g}^{\alpha\beta} W^\mu_{\ \alpha\beta,\nu} - \tfrac{1}{5}(3\lambda_1 + 1)(\mathfrak{g}^{\alpha\mu} W^\beta_{\ \alpha\beta,\nu} + \mathfrak{g}^{\mu\alpha} W^\beta_{\ \beta\alpha,\nu})$$
$$- \tfrac{2}{15}(5\lambda_2 - 9\lambda_1 + 2)\mathfrak{g}^{[\alpha\mu]} W_{\alpha,\nu}. \tag{22}$$

Finally, let us record the nonsymmetric analogues (there are several) of the equation of geodesic deviation. Again, in GR, it is this equation rather than the geodesic itself that represents an equation of motion if we consider two freely falling or inertial observers communicating their findings to each other. Strictly speaking it is meaningless to talk in GFT of geodesic deviation except in the background Riemannian manifold of the theory which remains to be determined. Nevertheless, the technique by which the equation is obtained in GR, is perfectly applicable in spite of its geometrical interpretation being uncertain. I write down the results because they may throw some light on what should be regarded as the correct expression for the force.

We start as usual with a two-parameter family of surfaces in the manifold (the space–time which must exist if we are to have any physics especially in the macroscopic picture of reality). The family is given by

$$x^\lambda = x^\lambda(u, v), \tag{23}$$

where u and v are real scalar parameters. Whatever the geometry may be, or rather however it may be determined from the field equations, we can define the vectors

$$t^\lambda = \frac{\partial x^\lambda}{\partial u}, \quad n^\lambda = \frac{\partial x^\lambda}{\partial v}, \quad Q^\lambda = \frac{\partial t^\lambda}{\partial u} + \Gamma^\lambda{}_{\mu\nu} t^\mu t^\nu = \frac{\partial t^\lambda}{\partial u} + \Gamma^\lambda{}_{(\mu\nu)} t^\mu t^\nu. \tag{24}$$

We also have invariant operators

$$\mathbf{D}^+{}_u n^\lambda = \frac{\partial n^\lambda}{\partial u} + \Gamma^\lambda{}_{\mu\nu} n^\mu t^\nu; \quad \mathbf{D}^-{}_u n^\lambda = \frac{\partial n^\lambda}{\partial u} + \Gamma^\lambda{}_{\mu\nu} n^\nu t^\mu;$$

$$\nabla^+{}_v Q^\lambda = \frac{\partial Q^\lambda}{\partial v} + \Gamma^\lambda{}_{\mu\nu} Q^\mu n^\nu; \quad \nabla^-{}_v Q^\lambda = \frac{\partial Q^\lambda}{\partial v} + \Gamma^\lambda{}_{\mu\nu} Q^\nu n^\mu. \tag{25}$$

Then, since by assumed continuity

$$\partial t^\lambda / \partial v = \partial n^\lambda / \partial u, \tag{26}$$

we find that

$$\overset{+\;-}{\mathbf{D}}{}^2{}_u n^\lambda = \nabla^+{}_v Q^\lambda + R^\lambda{}_{\mu\nu\sigma} t^\mu t^\nu n^\sigma, \tag{27}$$

together with

$$\overset{+\;+}{\mathbf{D}}{}^2{}_u n^\lambda = \nabla^-{}_v Q^\lambda + 2\Gamma^\lambda{}_{[\mu\nu]} (\mathbf{D}^+{}_u n^\mu) t^\nu$$

$$+ \left(R^\lambda{}_{\mu\nu\sigma} + \overset{\lambda}{\underset{++}{\Gamma^+_{[\sigma\mu];\,\nu}}} + \overset{\lambda}{\underset{++}{\Gamma^+_{[\sigma\nu];\,\mu}}} \right) t^\mu t^\nu n^\sigma,$$

$$\overset{-\;+}{\mathbf{D}}{}^2{}_u n^\lambda = \nabla^-{}_v Q^\lambda + \left(R^\lambda{}_{\mu\nu\sigma} + \overset{\lambda}{\underset{++}{\Gamma^+_{[\sigma\mu];\,\nu}}} + \overset{\lambda}{\underset{++}{\Gamma^+_{[\sigma\nu];\,\mu}}} \right) t^\mu t^\nu n^\sigma,$$

$$\overset{-\;-}{\mathbf{D}}{}^2{}_u n^\lambda = \nabla^+{}_v Q^\lambda + 2\Gamma^\lambda{}_{[\mu\nu]} (\mathbf{D}^+{}_u n^\nu) t^\mu$$

$$+ (R^\lambda{}_{\mu\nu\sigma} + 2\Gamma^\rho{}_{[\nu\sigma]} \Gamma^\lambda{}_{[\mu\rho]} + 2\Gamma^\rho{}_{[\mu\sigma]} \Gamma^\lambda{}_{[\nu\rho]}) t^\mu t^\nu n^\sigma.$$

These are the desired formulae. As I have pointed out however, their interpretation must await further development of the theory.

7. The energy–momentum tensor in GFT

The contracted Bianchi identities (6.17) together with the uniqueness of the Einstein tensor in a Riemannian space–time under straightforward linearity assumptions are used in GR as a justification for writing

$$G_{\mu\nu} = -\kappa T_{\mu\nu}. \tag{1}$$

As we have seen, Einstein based his investigation of the nonsymmetric unified field theory on a criticism of this equation and the WFE (3.17) do not contain an energy–momentum tensor as required. Nevertheless, $T_{\mu\nu}$ is needed in any invariant theory of physics. It is simply there whether we like it or not, unless we are dealing only with what are commonly known as ghost fields. Therefore, it is not a problem of GFT whether $T_{\mu\nu}$ exists but how to calculate it.

Einstein and Kaufman (in their version of the generalised field theory) toyed with the idea that $\mathfrak{T}^{\mu}{}_{\nu}$ introduced in the preceding section might represent an energy–momentum tensor (density) observing however the obvious fact that it is a tensor density only for special relativistic coordinate transformations. There are other difficulties with regarding $\mathfrak{T}^{\mu}{}_{\nu}$ as energy–momentum.

It is clearly not related to the quasi-Bianchi identities (6.16) but this is not such a serious problem although one might expect the relationship to hold by analogy with general relativity. What is perhaps more important is that whatever (metric) tensor we may use for raising and lowering of indices

$$\mathfrak{T}^{\mu\nu} = a^{\mu\sigma}\mathfrak{T}^{\nu}{}_{\sigma}, \tag{2}$$

is not in general symmetric. The condition of its symmetry is in fact a severe restriction on the geometry of space–time and yet we are accustomed to having a symmetric energy–momentum tensor for a wide variety of gravitational and electromagnetic fields. It would be curious if interaction envisaged in GFT involved breaking of this condition. There is something worse. I have mentioned (1.5.6) a solution of the WFE which will be derived later. We can readily calculate $W^{\lambda}{}_{\mu\nu}$ corresponding to this solution only to find that then

$$\mathfrak{T}^{0}{}_{0} \equiv 0. \tag{3}$$

However, this component is invariably identified as mass–energy density and should not vanish for the most straightforward field whatever its interpretation.

There are perhaps already sufficient reasons for regarding $\mathfrak{T}^{\mu}{}_{\nu}$ as a spurious quantity. Moreover, if we anticipate the metric hypothesis of the next chapter (referred to in (1.5.11)) a solution of the problem of matter readily suggests itself.

Indeed, with the symmetric part of the connection $\tilde{\Gamma}^{\lambda}{}_{\mu\nu}$ given by a Christoffel bracket (constructed from a nonsingular, symmetric $(0, 2)$ tensor) we can construct immediately the corresponding (symmetric) Ricci tensor

$$\tilde{R}^{s}{}_{\mu\nu} = -\tilde{\Gamma}^{\sigma}{}_{(\mu\nu),\sigma} + \tilde{\Gamma}^{\sigma}{}_{(\mu\sigma),\nu} + \tilde{\Gamma}^{\sigma}{}_{(\mu\rho)}\tilde{\Gamma}^{\rho}{}_{(\sigma\nu)} - \tilde{\Gamma}^{\sigma}{}_{(\mu\nu)}\tilde{\Gamma}^{\rho}{}_{(\sigma\rho)}, \tag{4}$$

and the Einstein tensor

$$\tilde{G}_{\mu\nu} = \tilde{R}^{s}{}_{\mu\nu} - \tfrac{1}{2}a_{\mu\nu}(a^{\alpha\beta}\tilde{R}^{s}{}_{\alpha\beta} - 2\Lambda). \tag{5}$$

There is now no reason why we should not define

$$T_{\mu\nu} \stackrel{\text{Df}}{=} -(1/\kappa)\tilde{G}_{\mu\nu}. \tag{6}$$

Because (metric hypothesis)

$$\tilde{\Gamma}^{\lambda}_{(\mu\nu)} = \left\{ \begin{matrix} \lambda \\ \mu\nu \end{matrix} \right\}_a,$$

the tensor (6) will automatically satisfy the conservation law

$$T^{\mu\nu}{}^{00}{}_{;\nu}(\tilde{\Gamma}) = 0, \tag{7}$$

and it is a small price to pay that it can not be readily derived from the variational principle which led to the field equations of the theory. After all, the latter was constructed so as to avoid having to introduce $T_{\mu\nu}$ into its formal structure.

3 The Riemannian background of the generalised field theory

1. Bifurcation of geometry and physics: the weak principle of geometrisation

The principle of equivalence, except in its strongest and almost 'occult' form, does not by itself determine the geometry of the space–time in general relativity. Neither does the observation that the latter is curved although Einstein's fame may have rested on it. The scheme is fixed by the specific assumption that Riemannian space–time is a sufficient model of the world, or rather, of the gravitational field. And geometry is determined directly by the solution of the field equations for a given physical situation embodied in the energy–momentum tensor. This solution is an expression of both the gravitational potential and the symmetric metric of the V_4, from which its whole structure can be derived by well-known rules. Thus the field equations (and the nonrelativistically defined $T_{\mu\nu}$) are all that we need to know in GR.

Does this situation carry over to GFT? The answer is emphatically no, or at least not so far. Let us review briefly what we have.

We started with an unspecified, unrestricted and therefore not symmetric affine connection $\Gamma^{\lambda}{}_{\mu\nu}$ and, possibly only partial since Γ_{μ} looks alluringly physical, an expression for the physical field (presumably the total field of Einstein) in the nonsymmetric tensor $g_{\mu\nu}$. We derived the unique field equations (WFE, 2.3.17) by an application of the principle of Hermitian symmetry, and a tacit use of Occam's razor in selecting the variational action principle (2.3.6.). The first of the equations (2.3.17a):

$$g_{\mu\nu,\lambda} - \tilde{\Gamma}^{\sigma}{}_{\mu\lambda}g_{\sigma\nu} - \tilde{\Gamma}^{\sigma}{}_{\lambda\nu}g_{\mu\sigma} = 0,$$

is well known (and as will be shown explicitly when we are discussing solutions of the field equations) to admit a unique solution for $\tilde{\Gamma}^{\lambda}{}_{\mu\nu}$ in terms of $g_{\mu\nu}$ and its derivatives, provided only that

$$g(g - 2) \neq 0. \tag{1}$$

This result was obtained independently by Mme Tonnelat and by

Hlavaty. But the relation (2.3.16)

$$\tilde{\Gamma}^{\lambda}{}_{\mu\nu} = \Gamma^{\lambda}{}_{\mu\nu} + \tfrac{2}{3}\delta^{\lambda}{}_{\mu}\Gamma_{\nu},$$

between Schrödinger's connection and the original $\Gamma^{\lambda}{}_{\mu\nu}$ can not be solved for the latter because its rank as an algebraic equation is 60 instead of 64 because

$$\tilde{\Gamma}_{\mu} \equiv 0.$$

This alone reinforces the possibility that the vector Γ_{μ} may have more to do with physics than with geometry except as an ephemeral part of the affine connection.

What is more important is that PHS which anchors the theory in a physical background, tells us nothing about the geometry of the space–time. The connection between physics and geometry, which, since the theory should relate to general relativity, must not be irretrievably broken, will have to be established by a further assumption or postulate.

The situation is no different than in GR with no more than the equality between gravitational and inertial mass as a heuristic hypothesis. Can the connecting postulate be similar to the GR case, so that this or that geometry is the model of the total field? I do not know what a nonsymmetric $g_{\mu\nu}$ would mean although, of course, a non-symmetric $\Gamma^{\lambda}{}_{\mu\nu}$ is perfectly admissible. And we know already from the paracompactness argument that space–time should admit a Riemannian metric. The question is how it should be determined if it is not given as an immediate solution of the field equations.

In the initial stages of the nonsymmetric theory Einstein had little choice but to assume that the metric

$$a_{\mu\nu} = g_{(\mu\nu)}. \tag{2}$$

Where does this leave $g_{[\mu\nu]}$? Other metrics which were algebraically dependent on $g_{[\mu\nu]}$ were proposed, notably by Wyman and by Schrödinger. These however were *ad hoc* definitions ungrounded in any meaningful physical, or even geometrical, hypothesis (though perhaps Schrödinger's volume argument could at a stretch be given geometrical meaning). Clearly one could multiply such proposals freely. And as far as Einstein's choice (2) is concerned, I shall show presently that it is incompatible with PHS.

It follows now that GFT necessarily leads to an initial bifurcation of geometry and physics. This, of course, was already contained in the assertion that the former was represented by the affine connection $\Gamma^{\lambda}{}_{\mu\nu}$ and the latter by the tensor $g_{\mu\nu}$ (and perhaps the vector Γ_{μ}). On the other hand, the principle of reducibility requires that in some sense they should be fused together again. PR can not be avoided partly because of the startling empirical success of general relativity and partly because without it we should be plunged into a complete, conceptual and empirical darkness. Physics progresses in measured steps and even GR, revolutionary as it was, arose out of the inability of special relativity to deal with gravitation. There are also the more philosophical reasons of looking to GFT for an account of the limitations of the gravitational theory.

When then should the 'fusion' of geometry and physics occur? If the field equations of GFT do not give it directly, indeed, if they only emphasise their bifurcation, then we can conclude that the conceptual unity must be re-established in every particular case after the equations have been solved. In other words, given a solution of the field equations we need a prescription which will enable us to determine the geometrical structure of space–time. This is the concept embodied in the metric hypothesis introduced below.

The initial bifurcation of geometry and physics and the requirement that in spite of it they must determine each other by whatever the theory tells us is the content of what I called the weak principle of geometrisation. It replaces the strong principle of GR and we shall see that only in this form can it remain consistent with the more heuristic idea of Hermitian symmetry.

2.   The conditions on a metric

PHS is the guiding principle of the generalised field theory with which everything else must be at least consistent. But GFT is essentially an 'unknown' theory because to start with it operates in an empirical vacuum. In these circumstances it is important not to introduce unnecessary assumptions, that is, to work with as much freedom and generality as possible.

Now a metric tensor is in the first instance needed to raise or lower tensor indices (and therefore only by inference to define vector length). The formulae for the covariant differentiation of vectors ('parallel transfer') do not require it. For example, the formula for $\delta h_{\underset{+}{\lambda}}$ follows from

$$\delta \overset{\lambda}{h}{}^{+} = -\Gamma^{\lambda}{}_{\mu\nu} h^{\mu} \delta x^{\nu} \tag{1}$$

if we demand only that $h_{\lambda} h^{\lambda}$ should be a scalar, that the product rule should hold for the δ operation and that the covariant derivative of a scalar should be the ordinary partial derivative. The corresponding formulae for tensors of higher rank follow by treating them as outer products of vectors: for example,

$$\delta T^{-}{}_{\mu\,\nu}_{+\,0} \equiv \delta(h^{-}{}_{\mu} k^{\lambda}{}_{\mu} l_{\nu})_{+\,0}$$

$$= -\Gamma^{\lambda}{}_{\sigma\alpha} T^{\alpha}{}_{\mu\nu}\delta x^{\sigma} + \Gamma^{\alpha}{}_{\mu\sigma} T^{\lambda}{}_{\alpha\nu}\delta x^{\sigma} + \Gamma^{\alpha}{}_{(\nu\sigma)} T^{\lambda}{}_{\mu\alpha}\delta x^{\sigma}. \quad (2)$$

On the other hand, the field equation

$$g_{\mu\nu;\lambda}_{+\,-}(\tilde{\Gamma}) = 0,$$

is not the most general expression which is Hermitian symmetric in $g_{\mu\nu}$ and $\tilde{\Gamma}^{\lambda}{}_{\mu\nu}$. This expression is clearly

$$g_{\mu\nu;\lambda}_{+\,-} + p(\tilde{\Gamma}^{\sigma}{}_{[\mu\lambda]}g_{(\sigma\nu)} + \tilde{\Gamma}^{\sigma}{}_{[\lambda\nu]}g_{(\mu\sigma)})$$

$$+ q(\tilde{\Gamma}^{\sigma}{}_{[\mu\lambda]}g_{[\sigma\nu]} + \tilde{\Gamma}^{\sigma}{}_{[\lambda\nu]}g_{[\mu\sigma]}) + r(\Gamma_{\mu}g_{\lambda\nu} - \Gamma_{\nu}g_{\mu\lambda})$$

$$+ s(\Gamma_{\mu}g_{\nu\lambda} - \Gamma_{\nu}g_{\lambda\mu}) = 0 \quad (3)$$

where p, q, r, s are arbitrary numbers. If we do not consider such expressions any further it is only because the variational principle we have adopted (on the basis of the principle of simplicity or Occam's razor) led to the weak field equations instead. But it told us nothing about the metric.

The metric tensor itself determines the basic laws of measurement in a space–time and therefore has both a physical and a geometrical significance. And by the weak principle of geometrisation both the geometry and the physical laws relevant to a fundamental theory must be uniquely determined by its content. Hence the metric tensor must be a symmetric tensor depending on the symbolic description of the field. (Skew or spin metrics may determine observability of physical events but hardly a measure of distance.) We write therefore

$$a_{\mu\nu} = a_{\nu\mu} = a_{\mu\nu}(g_{\alpha\beta}, \tilde{\Gamma}^{\gamma}{}_{\alpha\beta}, \ldots),$$

and, of course, require $a_{\mu\nu}$ to be a.e. nonsingular in the manifold.

Similarly, the principle of reducibility requires that when

$$g_{[\mu\nu]} = 0 = \tilde{\Gamma}^{\lambda}{}_{[\mu\nu]} \quad (4)$$

$a_{\mu\nu}$ should become the general relativistic $g_{(\mu\nu)}$. This represents an implicit acceptance of GR as a valid theory of gravitation in the absence of any other fields such as, in particular, electromagnetism, and in regions away from material sources. But it is the only condition which we can so far impose.

3.   The local Poincaré group

Let us now recall the well-known fact that if the local invariance group in a space–time is the Lorentz group we obtain the field structure of general relativity. Hence GFT requires an extension of the local group. And again the principle of reducibility demands that this extension should be closely related to the Lorentz structure. This leads us to consider the Poincaré group of affine transformations $A(4; \mathbb{R})$. I am indebted for this idea to my student, C. J. Radford who worked out the expressions in local coordinates given below.

Let the space–time manifold be denoted by X and let x be a point in it

$$x \in X. \tag{1}$$

The group $A(4; \mathbb{R})$ acts on the bundle $A(X)$ of affine frames of reference. It can always be represented (see Kobayashi & Nomizu 1963) as the group of all matrices of the form

$$\begin{Bmatrix} g & \xi \\ 0 & 1 \end{Bmatrix} \tag{2}$$

where $g \in$ the general linear group $GL(4; \mathbb{R})$ and $\xi \in \mathbb{R}^4$. Let the tangent affine space at x be denoted by $TA_x(X)$ and let $\mathbb{R}^4$ be assigned a Minkowski inner product. An affine frame $(a; X_i)$ at $x \in X$, consists of an $a \in TA_x(X)$ together with a linear frame $(X_i)_x$. Here, as in the rest of this section, Latin indices denote frame components.

The group $A(4; \mathbb{R})$ can be written as a semi-direct product

$$A(4; \mathbb{R}) = GL(4; \mathbb{R}) \cdot \mathbb{R}^4 \tag{3}$$

to which there corresponds the semi-direct sum of the Lie algebras:

$$a(4; \mathbb{R}) = gl(4; \mathbb{R}) \oplus \mathbb{R}^4. \tag{4}$$

Our aim is to determine linear connections and curvature forms within this structure. A linear connection ω is a $gl(4; \mathbb{R})$-valued 1-form on X. Let ψ be an $\mathbb{R}^4$-valued 1-form (i.e. a tensor of type $(1, 1)$) on X. Then a connection $\tilde{\omega}$ in $A(X)$ is given by

$$\tilde{\omega} = \omega \oplus \psi. \tag{5}$$

The curvature form of $\tilde{\omega}$ is, as usual,

$$\tilde{\Omega} = d\tilde{\omega} + \tfrac{1}{2}[\tilde{\omega}, \tilde{\omega}], \tag{6}$$

where d is the exterior differentiation operator and $[\tilde{\omega}, \tilde{\omega}]$ is the Lie bracket in the simplified, mnemonic notation (so that it is not zero). We can split (6) into a $gl(4; \mathbb{R})$-valued

$$\Omega = d\omega + \tfrac{1}{2}[\omega, \omega], \tag{7}$$

and an $\mathbb{R}^4$-valued

$$\tilde{\Theta} = \mathrm{d}\psi + \tfrac{1}{2}[\psi, \omega] + \tfrac{1}{2}[\omega, \psi] = \mathrm{D}\psi, \tag{8}$$

D being the covariant derivative with respect to ω. When ψ is the canonical 1-form, the generalised torsion $\tilde{\Theta}$ becomes the standard torsion form.

We can now rewrite these standard results as follows in terms of local coordinates. Let $E^i{}_j$ be the natural basis of the algebra $gl(4; \mathbb{R})$; e_j, the natural basis in $\mathbb{R}^4$; y^α, the local coordinates in $TA_x(X)$; $X^\alpha{}_i$, the local coordinates in the bundle $L(X)$ of linear frames; and x^α, the local coordinates in X.

The term 'local' must be thought of as referring to a uniform neighbourhood of a point in the space–time manifold. The local coordinates in the bundle $A(X)$ are then $(y^\alpha; X^\alpha{}_i; x^\alpha)$ while $(y^\alpha; X^\alpha{}_i)$ are the corresponding local coordinates in the fiber $\pi^{-1}(x)$ where

$$\pi : A(X) \to X \tag{9}$$

is the projection mapping of the affine bundle into the space–time manifold. Then

$$\omega = E^i{}_j(Y^j{}_\alpha \mathrm{d}X^\alpha{}_i + \gamma^j{}_{\alpha i}\mathrm{d}x^\alpha), \tag{10}$$

and

$$\psi = e_j Y^j{}_\alpha(\mathrm{d}y^\alpha + K^\alpha{}_\beta \mathrm{d}x^\beta), \tag{11}$$

where $K^\alpha{}_\beta$ is a $(1, 1)$ tensor and $\gamma^j{}_{\alpha i}$ are the local components of the Ricci coefficients of rotation. Also

$$Y^j{}_\alpha = (X^{-1})^j{}_\alpha. \tag{12}$$

Let $X^*{}_\alpha$ be the horizontal lift of

$$X_\alpha = \partial_{x^\alpha},$$

so that

$$\tilde{\omega}(X^*{}_\alpha) = 0.$$

Then

$$X^*{}_\alpha = \partial_{x^\alpha} - \gamma^i{}_{\alpha j}X^\beta{}_i\partial_{x^\beta_j} - K^\beta{}_\alpha\partial_{y^\alpha}. \tag{13}$$

If now $a \in A_x(X)$ so that

$$\pi(a) = x \in X,$$

then with every vector V we can associate an A^4-valued function

$$f : f(a) = a^{-1}(V(\pi(a)))$$

when the function associated with X_α will be just

$$f_\alpha = a^{-1}(X_\alpha) = Y^j{}_\alpha e_j.$$

The A^4-valued function associated with the covariant derivative $\nabla_{x^\alpha} X_\beta$, is now

$$X^*_\alpha(f_\beta) = (X^\gamma_j Y^j_{\beta,\alpha} + \gamma^k_{\alpha j} X^\gamma_k Y^j_\beta) f_\gamma, \tag{14}$$

and we can write

$$\nabla_{\partial x^\alpha} \partial X_\beta = \Gamma^\gamma_{\alpha\beta} \partial_{x^\gamma}, \tag{15}$$

where the nonsymmetric connection $\Gamma^\gamma_{\alpha\beta}$ is given by

$$\Gamma^\gamma_{\alpha\beta} = X^\gamma_j Y^j_{\beta,\alpha} + \gamma^k_{\alpha j} X^\gamma_k Y^j_\beta. \tag{16}$$

This represents the connection coefficient with respect to ω. It follows immediately from the definition (16) that

$$\nabla_\alpha X^\beta_j = X^\beta_{j,\alpha} + \Gamma^\beta_{\alpha\sigma} X^\sigma_j - \gamma^k_{\alpha j} X^\beta_k = 0. \tag{17}$$

Now the elements of the tangent bundle $T(X)$ of the space–time manifold have the same structure as the real vector space $\mathbb{R}^4$ with the Minkowski product. Hence, reducing the group GL $(4;\mathbb{R})$ to the Lorentz group restricts the linear bundle $L(X)$ to the Lorentz bundle $\mathscr{L}(X,L)$. Transformations of the latter preserve the tensor

$$l^{\alpha\beta} = X^\alpha_i \eta^{ij} X^\beta_j, \tag{18}$$

when the Minkowski tensor η^{ij} becomes the fiber metric. Similar restriction of the affine to Poincaré transformations, reduces $A(X)$ to the Poincaré bundle of frames, say, $P(X)$.

If we raise and lower Latin indices with η^{ij}, two conditions must be satisfied by whatever we may choose as the space–time metric. First the Ricci coefficients must be skew

$$\gamma_{\alpha ij} = -\gamma_{\alpha ji}, \tag{19}$$

and secondly, the selected definition must not violate the principle of Hermitian symmetry.

4.   The metric hypothesis of GFT

It turns out that we are not free to choose the space–time metric at will. Indeed, the second of the above conditions, that in whatever way we may define the metric of space–time, the definition must not violate PHS, excludes $l^{\alpha\beta}$ as a possible metric tensor. This means that the latter can not be constructed from the tetrad. Instead, it must be constructed from a new tetrad h^α_i:

$$a^{\alpha\beta} = h^\alpha_i \eta^{ij} h^\beta_j \tag{1}$$

which does not satisfy equation (3.17) but an equation

$$\nabla_\gamma h^\alpha_i = \tilde{\Gamma}^\alpha_{[\gamma\sigma]} h^\sigma_i. \tag{2}$$

This definition (of $h^\alpha{}_i$) is unique in view of the field equations (2.3.17). I should point out that this is not a circular argument in GFT although it would be in general relativity in which the components of the metric tensor are identified with the gravitational potentials or with the physical interpretation of the theory. The weak field equations have been derived independently of the metric definition. All that we are compelled to require is that this definition should not contradict the principles of the theory. Hence uniqueness of the equation (2) is guaranteed, since $\tilde{\Gamma}^\lambda{}_{\mu\nu}$ is the connection (albeit auxilliary) of GFT.

With the tetrad defined by equation (2),

$$D_\gamma h^\alpha{}_i \overset{\text{Df}}{=} h^\alpha{}_{i,\gamma} + \tilde{\Gamma}^\alpha{}_{(\gamma\sigma)} h^\sigma{}_i - \gamma^k{}_{\gamma i} h^\alpha{}_k = 0. \tag{3}$$

Also

$$D_\gamma \eta^{ij} = \nabla_\gamma \eta^{ij} = 0, \tag{4}$$

and it follows that

$$D_\gamma a^{\alpha\beta} = D_\gamma a_{\alpha\beta} = a_{\alpha\beta,\gamma} - \tilde{\Gamma}^\sigma{}_{(\alpha\gamma)} a_{\sigma\beta} - \tilde{\Gamma}^\sigma{}_{(\beta\gamma)} a_{\alpha\sigma} = 0. \tag{5}$$

It can be easily verified that equation (5) which, as long as the tetrad $h^\alpha{}_i$ is not null, defines the nonsingular metric $a_{\alpha\beta}$, is also a necessary and sufficient condition for the skewsymmetry of the Ricci rotation coefficients (3.19). Another form of equation (5) is

$$\left\{ \begin{matrix} \lambda \\ \mu\nu \end{matrix} \right\}_a = \tilde{\Gamma}^\lambda{}_{(\mu\nu)}, \tag{6}$$

or the requirement that the symmetric part of the Schrödinger affine connection should be a Christoffel bracket formed from the metric tensor of the theory. It was in this form that the definition was originally guessed as a general relation satisfying what now appears as an initial condition that

$$a_{\mu\nu} = g_{(\mu\nu)}(\tilde{\Gamma}^\gamma{}_{(\alpha\beta)}).$$

I have called equation (6) the metric hypothesis in my published articles. When it was first introduced, it was indeed a hypothesis although the discussion of the last two sections shows that it is a necessary condition of compatibility of the GFT principles, a theorem rather than an assumption.

We must now consider the consistency of the metric hypothesis (as I shall continue to refer to it). Equation (5) is a set of differential equations for the components of the metric tensor. Although it represents in general forty equations for only ten unknowns the system is formally integrable. Indeed, the integrability condition

$$a_{\mu\nu,\alpha\beta} = a_{\mu\nu,\beta\alpha} \tag{7}$$

is equivalent to

$$\tilde{R}^\sigma{}_{\mu\lambda\kappa} a_{\sigma\nu} = \tilde{R}^\sigma{}_{\nu\kappa\lambda} a_{\mu\sigma}, \tag{8}$$

where $\tilde{R}^\lambda{}_{\mu\nu\kappa}$ is the Riemann–Christoffel tensor constructed from the connection $\tilde{\Gamma}^\lambda{}_{(\mu\nu)}$. But because of the metric hypothesis, this is also the curvature tensor of a Riemannian V_4 whose metric is $a_{\mu\nu}$. Hence equation (8) is equivalent to the well-known symmetry relation

$$\tilde{R}_{\nu\mu\lambda\kappa} = \tilde{R}_{\mu\nu\kappa\lambda}, \tag{9}$$

of any V_4. This proves the integrability assertion.

Then, uniqueness of the 'general relativistic' solution of equation (2) defining the metric tetrad is assured given the reduction to $g_{(\mu\nu)}$, say, at a fixed point of the manifold. This result is discussed in detail, for example by Spivak (vol. 1, ch. 6) and we need not repeat it here.

Let us now write

$$\tilde{\Gamma}^k{}_{\alpha j} = h^k{}_\sigma h^\beta{}_j \tilde{\Gamma}^\sigma{}_{(\alpha\beta)} \quad \text{and} \quad \Gamma^k{}_{\alpha j} = X^k{}_\sigma X^\beta{}_j \tilde{\Gamma}^\sigma{}_{(\alpha\beta)}. \tag{10}$$

Then the rotation coefficients

$$h^k{}_\sigma h^\sigma{}_{j,\alpha} + \tilde{\Gamma}^k{}_{\alpha j} \quad \text{and} \quad X^k{}_\sigma X^\sigma{}_{j,\alpha} + \Gamma^k{}_{\alpha j}, \tag{11}$$

give 1-forms in two distinct subbundles in the affine bundle $A(X)$. The subbundles must then be related by a transformation of the form

$$\omega' g = g\omega + \mathrm{d}g, \tag{12}$$

reverting for a moment to the index-free notation. On the other hand, Ricci coefficients of rotation are defined in (3.10) for the whole affine bundle $A(X)$. Hence we must have

$$\omega' = \omega \tag{13}$$

so that the transformation equation (12) becomes an exterior differential equation

$$\mathrm{d}g = \omega g - g\omega \tag{14}$$

for the transformation g. The integrability condition of this equation is

$$\Omega g = g\Omega \tag{15}$$

where Ω is the curvature form of the manifold. This can be nontrivially ($g \neq$ constant) satisfied by, for example,

$$g = \Omega \tag{16}$$

(it does not matter that apparently g and Ω are of different nature; we consider (15) as a matrix equation and this allows (16) to be written down). However, since Ω is skew, g must in general belong to the symplectic group. Existence of such a nontrivial solution now ensures that equation (13) can be written down without unduly restricting either the affine bundle, or, what would be worse, the manifold itself.

5. The weak principle of geometrisation revisited

The group theoretical argument must be developed a little further. First, the basis of the Lorentz group is

$$L_{kj} = E_{[kj]}, \tag{1}$$

in terms of the natural basis

$$E^i{}_j = \eta^{ik} E_{kj} \tag{2}$$

The Poincaré group is then generated by $\{L_{ij}, l_k\}$ and its Lie brackets are

$$\left.\begin{array}{l} [L_{ij}, L_{kl}] = \eta_{jk} L_{il} - \eta_{ik} L_{jl} + \eta_{jl} L_{ki} - \eta_{li} L_{kj}, \\ [L_{ij}, l_k] = \eta_{jk} l_i - \eta_{ki} l_j. \end{array}\right\} \tag{3}$$

Let σ be the cross-section of $P(X)$ over an open neighbourhood U of a point x of the manifold. To every $x \in U$, σ assigns a Poincaré frame $(O_x, X^\alpha{}_i \partial_{x^i})$, O being the origin of $\mathbb{R}^4$. In terms of the coordinates $(y^\alpha, x^\alpha, X^\alpha{}_i)$ of the affine bundle $A(X)$, σ is the mapping

$$\sigma : (x^\alpha) \to ((0,0,0,0), x^\alpha, X^\alpha{}_i). \tag{4}$$

We then have on the base manifold X the following relations

$$\left.\begin{array}{l} \sigma^* \omega^i{}_j = \gamma^i{}_{\alpha j} dx^\alpha, \\ \sigma^* \phi^j = K^j{}_\alpha dx^\alpha, \\ \sigma^* \Omega^i{}_j = \tfrac{1}{2} R^i{}_{j\mu\nu} dx^\mu \wedge dx^\nu, \end{array}\right\} \tag{5}$$

and

$$\left.\sigma^* \bar{\Theta}^j = \tfrac{1}{2} \bar{T}^j{}_{\mu\nu} dx^\mu \wedge dx^\nu, \right\}$$

where

$$R^i{}_{j\mu\nu} = \gamma^i{}_{\nu j,\mu} - \gamma^i{}_{\mu j,\nu} + \gamma^k{}_{\nu j} \gamma^i{}_{\mu k} - \gamma^k{}_{\mu j} \gamma^i{}_{\nu k}, \tag{6}$$

and

$$\bar{T}^j{}_{\mu\nu} = 2[\nabla_{[\mu} K^j{}_{\nu]} + \tilde{\Gamma}^\sigma{}_{[\mu\nu]} K^j{}_\sigma]. \tag{7}$$

The star in equations (5) denotes the 'pull through' product and not, of course, the operation of Hermitian conjugation.

If the two tetrads coincide (as they must not if the affine bundle is to split in the required way):

$$K^j{}_\nu = h^j{}_\nu = X^j{}_\nu, \tag{8}$$

$\bar{T}^j{}_{\mu\nu}$ becomes the torsion tensor of the affine connection $\tilde{\Gamma}^\lambda{}_{\mu\nu}$. As it is, we shall call it generalised torsion. However, without the forbidden restriction (8),

$$R^\alpha{}_{\beta\mu\nu} = X^\alpha{}_i X^j{}_\beta R^i{}_{j\mu\nu} \tag{9}$$

is the Riemann–Christoffel tensor constructed from the connection $\tilde{\Gamma}^\lambda{}_{\mu\nu}$. Let us define the (1, 1) tensor

$$K^\alpha{}_\mu = h^\alpha{}_j K^j{}_\mu. \tag{10}$$

Then, raising and lowering Greek indices with the metric tensor $a_{\mu\nu}$, we find that the generalised torsion is

$$\bar{T}^j{}_{\mu\nu} = h^{j\sigma}[\nabla_\mu K_{\sigma\nu} - \nabla_\nu K_{\sigma\mu} + \tilde{\Gamma}^\alpha{}_{[\mu\sigma]} K_{\alpha\nu} - \tilde{\Gamma}^\alpha{}_{[\nu\sigma]} K_{\alpha\mu}]. \tag{11}$$

If we therefore identify $K_{\mu\nu}$ with the fundamental tensor $g_{\mu\nu}$ which thus acquires quasi-geometrical meaning as well as representing physical fields, and for which

$$\nabla_\lambda g_{\mu\nu} = g_{\mu\nu, \lambda} - \tilde{\Gamma}^\sigma{}_{\mu\lambda} g_{\sigma\nu} - \tilde{\Gamma}^\sigma{}_{\lambda\nu} g_{\mu\sigma} = 0,$$

the generalised torsion becomes

$$\bar{T}^j{}_{\mu\nu} = h^{j\sigma} g_{\alpha\beta} [\tilde{\Gamma}^\alpha{}_{[\nu\sigma]} \delta^\beta{}_\mu - \tilde{\Gamma}^\alpha{}_{[\mu\sigma]} \delta^\beta{}_\nu]. \tag{12}$$

The identification

$$K_{\mu\nu} = g_{\mu\nu} \tag{13}$$

is virtually mandatory. We do not want too many unidentifiable tensors floating about the theory, in fact we do not want any, and the variationally derived field equations have been formulated in terms of $g_{\mu\nu}$ and $\tilde{\Gamma}^\lambda{}_{\mu\nu}$ only. We may note also that torsion can not vanish. The condition for $\bar{T}^j{}_{\mu\nu}$ to be zero is

$$g^{[\mu\nu]} \tilde{\Gamma}^\sigma{}_{[\mu\nu]} = 0, \tag{14}$$

and this would be too strong a restriction on the field.

Since it is the physical objects, that is the $g_{\mu\nu}$ field, that are expressed in terms of the h-tetrad, we can call it the physical frame. The affine connection and the Riemann–Christoffel tensor being expressed in terms of the X-tetrad, we can similarly refer to it as geometrical. Thus, the two-frame formulation of the group-theoretical background of the nonsymmetric theory is a restatement of the weak principle of geometrisation.

I have mentioned before that Einstein's identification of the metric (1.2)

$$a_{\mu\nu} = g_{(\mu\nu)}$$

is incompatible with the principle of Hermitian symmetry. Actually it is incompatible with the group-theoretic or fiber-bundle formulation of

GFT in terms of the two unequal tetrads, which of course, is designed to ensure compatibility with PHS. To see this, let us write, as is often convenient,

$$g_{(\mu\nu)} = h_{\mu\nu} \quad \text{and} \quad g_{[\mu\nu]} = k_{\mu\nu}. \tag{15}$$

Then

$$h_{\mu\nu,\lambda} - \tilde{\Gamma}^{\sigma}{}_{(\mu\lambda)}h_{\sigma\nu} - \tilde{\Gamma}^{\sigma}{}_{(\lambda\nu)}h_{\mu\sigma} = \tilde{\Gamma}^{\sigma}{}_{[\mu\lambda]}k_{\sigma\nu} + \tilde{\Gamma}^{\sigma}{}_{[\lambda\nu]}k_{\mu\sigma}. \tag{16}$$

Our formulation of the theory, however, requires that the metric should satisfy either an equation of the form

$$\nabla_{\gamma}a_{\alpha\beta} = 0 \tag{17}$$

(which I have rejected) or

$$D_{\gamma}a_{\alpha\beta} = 0,$$

which is the metric hypothesis (4.5). Equation (1.2) would therefore be possible only if

$$\tilde{\Gamma}^{\sigma}{}_{[\mu\lambda]}k_{\sigma\nu} + \tilde{\Gamma}^{\sigma}{}_{[\lambda\nu]}k_{\mu\sigma} = 0, \tag{18}$$

a restriction which is both physically and geometrically meaningless. We are now in possession of the complete structure of the non-symmetric theory. I shall close this chapter with a brief discussion of the significance of adjoining the definition of a Riemannian metric by the equations (4.5) or (4.6) to the set of the weak field equations (2.3.17).

6. Significance of the metric hypothesis

The generalised field theory is given by the equations

$$\left.\begin{array}{l} g_{\mu\nu;\lambda}(\tilde{\Gamma}) = 0; \quad \mathrm{g}^{[\mu\nu]}{}_{,\nu} = 0; \quad R_{(\mu\nu)}(\tilde{\Gamma}) = 0; \\[4pt] R_{[\mu\nu]}(\tilde{\Gamma}) = \tfrac{2}{3}(\Gamma_{\mu,\nu} - \Gamma_{\nu,\mu}); \quad \tilde{\Gamma}^{\lambda}{}_{\mu\nu} = \Gamma^{\lambda}{}_{\mu\nu} + \tfrac{2}{3}\delta^{\lambda}{}_{\mu}\Gamma_{\nu}; \\[4pt] \begin{Bmatrix} \lambda \\ \mu\nu \end{Bmatrix}_{a} = \tilde{\Gamma}^{\lambda}{}_{(\mu\nu)}. \end{array}\right\} \tag{1}$$

They are a unique set of equations which follow from the principle of Hermitian symmetry and the group-theoretical foundations outlined in the preceding sections. Since $\tilde{\Gamma}^{\lambda}{}_{\mu\nu}$ is uniquely determined from the first of the equations (1) and the metric $a_{\mu\nu}$ from the last, we can say that a solution of equations (1) determines both the physical content and the geometry of the world. In this sense, the concept of geometrisation of physics is retained in GFT as required, although in a different way than in general relativity.

Of course, we must still demonstrate that the field equations can be solved, interpret the solution physically and show that any conclusions that can be drawn are not in conflict with empirical evidence or for that matter with what is regarded as already known about the physical world. Two observations, however, can be made before we proceed to these more practical aspects of GFT.

The first is that the metric hypothesis (the last of equations (1)) must limit the range of possible solutions of the field equations. This follows from the fact that, as in GR, the field equations can be solved in practice only under severe symmetry restrictions. And it can not be ascertained that a solution for $a_{\mu\nu}$ will exist in all cases. Those cases for which the metric can not be determined, will have to be rejected as physically meaningless. In other words, the metric hypothesis can be regarded as having the status of a new 'law of physics'.

But as a law its nature is essentially geometrical. It determines the Riemannian metric of the space–time in the presence of interacting, macrophysical fields. We have still to show that these are gravitation and electromagnetism although reducibility of GFT to GR when symmetry is reimposed already indicates that at least the pure gravitational field will be described by the theory as required.

There is another aspect which follows from our structure and which was not envisaged by Einstein when he was first constructing his theory because at that time the view was still prevalent that an extension of Riemannian geometry was needed for a unified description of the total field. This was an inescapable consequence of the strong principle of geometrisation. However, relaxation of the principle, the metric hypothesis and a closer investigation of the nature of physical space–time show that its geometry must be Riemannian and that relations in a V_4 still describe the fundamental laws of physical measurement. What has changed is not the background geometry but the way in which it is to be determined by the laws of the total field summarised in the equations (1). This then is the basic difference between the generalised field theory and general relativity.

7.   A plan of action

I have now completed the account of the structure of GFT. The assumptions and principles of the theory have been formulated, the field equations implied by them derived, and uniqueness of the equations discussed. Since the latter are obtained from a self-consistent variational principle, their formal integrability is assured. However, a theory claiming to be physical must lead to physically meaningful results and must be interpreted in relation to empirically measurable quantities while so far, we have only an empty mathematical scheme, whatever its consistency with apparently physical hypotheses may be.

We have attempted a weak field approximation (chapter 2, section 4) which was sufficient to identify gravitational potential in GR because of the principle of equivalence. But this principle has now been replaced by Hermitian symmetry and the weak principle of geometrisation which do not have such interpretative power. Moreover, in the case of weak fields, the supposed gravitation and electromagnetism (we do not even know as yet that these are the fields GFT talks about, although perhaps the principle of reducibility ensures that we at least have the former) decouple. The weak field equations also do not lead to Maxwell's theory but to the weaker equations (2.4.7):

$$\Box k_{[\mu\nu,\alpha]} = 0.$$

This is what we might have expected but it implies that we must look for an interpretation of the theory elsewhere. There are two avenues of investigation open to us at this stage.

Clearly, if the weak field approximation failed to provide enough information, we should see what exact solutions of the field equations may teach us. Nevertheless, there are drawbacks in seeking them at once. The story of general relativity shows how difficult exact solutions are to obtain. We may feel, too, that we know how to recognise mass in a given solution but this may not be sufficient to recognise charge. Indeed, it was an overhasty (though, of course, both excusable and perhaps inevitable) attempt to identify electromagnetic fields that got the nonsymmetric theory into severe problems. I am referring to the results of Infeld (1950) and Callaway (1953) which seemed to show that the field equations of GFT do not lead to the equations of motion of a charged test particle in an electromagnetic field. The particle moved as if there was no charge on it. There was no Lorentz force in the order of approximation in which it should have appeared.

I have tried in this monograph to follow as closely as possible a logical development of GFT which, if we leave aside the otherwise all important aspect of an empirical verification, would ensure what Einstein called its emergence as a 'natural' extension of GR but I prefer to call its 'reasonableness'. In particular, I tried to assume as little as possible while endowing the theory with its whole generality. The process of interpretation will necessarily involve us in making an additional assumption. But it should enhance the strength of the theory if it can be shown that this very process removes a fatal obstacle to the theory's acceptance.

I shall therefore tackle first the problem of motion. The equations describing the motion of a charged test particle will be derived using the approximation technique devised by Einstein, Infeld and Hoffman (EIH in 1938 to solve the corresponding problem in GR). Since the method is complicated, I shall start the next chapter, which is devoted to the problem of motion, by outlining its essential features before showing that the problem of motion in GFT can be resolved provided the electromagnetic field is suitably identified.

The reason why field equations can determine motion of particles is their nonlinearity but as Einstein pointed out frequently, this is only a necessary condition for the possibility. If the field equations were linear, a sum of two solutions corresponding to the world lines of, say, two particles would be a solution, so that interaction between them could not be determined. As a matter of fact a not incomparable difficulty arises in GFT with respect to, so called, 'similarity' solutions. This will have to be resolved separately. It is the existence of four quasi-Bianchi identities given by the equations (2.6.16), that ensures, together with the basic nonlinearity, that the equations of motion are derivable from the field equations while freedom of choice of the coordinate system is preserved. This is important because calculations of the EIH can be considerably simplified by imposing certain coordinate conditions.

Two things should be pointed out. First, the EIH method is valid whether particles are represented by singularities as in actual calculations described in the next chapter, or by singularity-free solutions of the field equations as Einstein hoped they might be, although for mathematical reasons it is unlikely that this hope can be realised. Secondly, the identities

$$g^{[\mu\nu]}{}_{,\nu} = 0$$

of the weak field equations (or $\tilde{\Gamma}_\lambda = 0$) are of a different nature from the Bianchi identities which depend on the variational principle. The former are essentially algebraic relations which ensure (in part) the compatibility of WFE with the principle of Hermitian symmetry. They have nothing to do with coordinate freedom although, together with coordinate symmetry conditions which alone make the field equations solvable in particular cases, they may affect the form of a given solution.

Only when the problem of motion has been resolved, shall we turn to seeking exact solutions whose nature, at least in the case of static spherical symmetry (where alone a general solution is known), will lead us to considerations of cosmology.

4 The problem of motion

1. An outline of the Einstein–Infeld–Hoffman method

The field equations of a nonlinear theory must be solved if we are to show that they imply equations of motion, but a general solution is impossible and would be meaningless even if it were not. Hence an approximate technique has to be devised which is not that of weak fields since a linearised theory will not do.

I shall follow here the EIH technique as simplified by Einstein & Infeld in 1949 with reference to the general relativistic field equations

$$R_{\mu v} = 0 \tag{1}$$

in empty space, or rather, outside any field-producing sources. The extension to GFT is relatively straightforward and will be introduced later, when we come to integration of WFE. In the absence of an energy–momentum tensor (deliberate), particles must be represented by field-singularities at least when we want to write down a concrete solution. As mentioned before, however, it will be clear from the use to which we shall put these particular expressions, that they can also be represented by solutions which are singularity free throughout space–time (if such exist).

The underlying idea of the EIH method is an expansion of the field quantities in terms of a single sufficiently small (for any subsequent expansion to converge) parameter, which is associated with time say. Selection of time is arbitrary but clearly convenient (if not forced) because whatever else may be said, a time parameter is distinguished by the signature of the space–time.

Thus, Einstein put

$$x^{\mu} = (\lambda \tau, x^1, x^2, x^3). \tag{2}$$

Expansion of field quantities as power series in λ means that they are supposed to vary slowly with the time $x^0 = \lambda \tau$. In other words, we have a quasi-stationary situation. Partial differentiation, denoted as usual by a comma, will refer to the coordinate system (τ, x^k), that is

$$x^0 \rightarrow \lambda x^0. \tag{3}$$

Two simple devices facilitate the calculation. The first is to write the Riemannian metric tensor (we are considering GR)

$$g_{\mu\nu} = \eta_{\mu\nu} + h_{\mu\nu}, \quad g^{\mu\nu} = \eta^{\mu\nu} + h^{\mu\nu}, \tag{4}$$

so that expansions of 'expressions in $g_{\mu\nu}$ and its derivatives' will start with a nonzero power of λ. Secondly, since

$$R = g^{\mu\nu} R_{\mu\nu} = 0, \tag{5}$$

we can extract the quasi-linear part of the Einstein tensor (which is more convenient than the Ricci tensor because of the Bianchi identities):

$$^{L}G_{\mu\nu} = R_{\mu\nu} - \tfrac{1}{2}\eta_{\mu\nu}\eta^{\alpha\beta}R_{\alpha\beta}, \tag{6}$$

which suggests a further replacement

$$\gamma_{\mu\nu} = h_{\mu\nu} - \tfrac{1}{2}\eta_{\mu\nu}\eta^{\alpha\beta}h_{\alpha\beta}, \quad h_{\mu\nu} = \gamma_{\mu\nu} - \tfrac{1}{2}\eta_{\mu\nu}\eta^{\alpha\beta}\gamma_{\alpha\beta}. \tag{7}$$

The energy–momentum tensor for a dust cloud

$$T^{\mu\nu} = \rho \frac{\mathrm{d}x^{\mu}}{\mathrm{d}s}\frac{\mathrm{d}x^{\nu}}{\mathrm{d}s}, \tag{8}$$

now suggests that we should have

$$\gamma_{ij} \sim \lambda\gamma_{0j} \sim \lambda^{2}\gamma_{00}. \tag{9}$$

(Here $\sim$ indicates corresponding orders of magnitude.) In fact Einstein assumed in consequence, that

$$\gamma_{00} = \sum_{1}^{\infty}\lambda^{2p}\gamma_{00}; \quad \gamma_{0i} = \sum_{1}^{\infty}\lambda^{2p+1}\underset{2p+1}{\gamma_{0i}}; \quad \gamma_{ij} = \sum_{1}^{\infty}\lambda^{2p+2}\underset{2p+2}{\gamma_{ij}}, \tag{10}$$

inclusion of alternate powers being analogues to having half-retarded and half-advanced potentials in Maxwell's electrodynamics.

The EIH method now consists of surrounding each particle (assuming that the field considered corresponds to, say, n particles or singularities) separately by a closed 2-surface and integrating (the field equations) over it up to a finite order p of approximation in λ. If the field equations happen to be integrable for the lowest power of λ, and the assumption of their integrability up to λ^{p} implies integrability in λ^{p+1} the problem will be solved by simple induction provided the expansion converges. Einstein and Infeld show that integrability in λ^{p+1} can always be ensured by imposing certain allowable conditions. It is these conditions that give the equations of motion.

To be more precise, let us assume that the field equation ((1) or $G_{\mu\nu} = 0$) are satisfied up to the order indicated:

$$\underset{2p-2}{G_{00}} = 0, \quad \underset{2p-3}{G_{0i}} = 0, \quad \underset{2p-2}{G_{ij}} = 0, \tag{11}$$

and let us consider Bianchi identities

$$G^{\mu\nu}{}_{;\nu} = 0. \tag{12}$$

The 'time-like' or zero identity ($\mu = 0$) of the order $2p - 1$, is,

$$\underset{2p-1}{G_{0k,k}} = 0, \tag{13}$$

all other terms disappearing because of (11). If we also have, in addition to (11),

$$\underset{2p-1}{G_{0i}} = 0, \tag{14}$$

then

$$\underset{2p}{G_{ij,j}} = 0. \tag{15}$$

Let us now recall that by Gauss' and Stokes' theorems a 2-integral for any skewsymmetric function (tensorial or not and with as many other indices as we like):

$$\int F_{ij,j} n_i \, \mathrm{d}S = 0, \quad F_{ij} = -F_{ji}, \tag{16}$$

where n_i is the unit normal vector to the 2-surface, and is independent of the shape of S (so that we can henceforth exclude particles by spherical surfaces). The trick now is to extract from the field equations as many expressions as can be skewsymmetrised and reject the relevant integrals. Of course, we can also use ${}^{L}G_{\mu\nu}$ instead of $G_{\mu\nu}$. Let us write in fact,

$$L_{\mu\nu} + 2N_{\mu\nu} = -2{}^{L}G_{\mu\nu} = 0, \tag{17}$$

where, with summation,

$$L_{00} = -\gamma_{00,kk}; \quad L_{\mu k} = (\gamma_{\mu j,k} - \gamma_{\mu k,j} - \delta_{\mu k}\gamma_{ji,i} + \delta_{\mu j}\gamma_{ki,i})_{,j}$$

and $N_{\mu\nu}$ contains all other linear, as well as nonlinear, terms. Then (16) applies to $L_{\mu k}$, whence

$$N_{\mu k,k} = 0, \tag{18}$$

as well as

$$\int^{s} 2N_{\mu k} n_k \, \mathrm{d}S = 0, \tag{19}$$

where s denotes the integral over an sth sphere.

The process used in obtaining equations (13) and (15) now adjoins to the system of equations

$$\underset{2p-2}{L_{00}} + 2\underset{2p-2}{N_{00}} = 0, \tag{20a}$$

$$\underset{2p-1}{L_{0k}} + 2\underset{2p-1}{N_{0k}} = 0,\tag{20b}$$

$$\underset{2p}{L_{ij}} + 2\underset{2p}{N_{ij}} = 0,\tag{20c}$$

the conditions

$$\underset{2p-1}{N_{0k,k}} = 0 = \underset{2p}{N_{ij,j}}.\tag{21}$$

What now happens is the following. If the equations can be satisfied for $p = 2$ (the lowest exponent) they will be satisfied for all the finite exponents, so that

$$\int^s 2\underset{2p-1}{N_{0k}}n_k\,\mathrm{d}S = \int^s 2\underset{2p}{N_{0k}}n_k\,\mathrm{d}S = 0.\tag{22}$$

Since however the Ns are known up to this order from the field equations alone, the vanishing of the integrals (22) will be extra conditions which can only be satisfied by an additional 'trick'. We must check first the $p = 2$ case, when

$$\underset{2}{\gamma_{00,kk}} = 0, \quad -\underset{3}{\gamma_{0k,jj}} + \underset{3}{\gamma_{0j,kj}} = \underset{2}{\gamma_{00,0k}}\tag{23}$$

(from (20a) and (20b)). It is here that we choose the harmonic solution

$$\underset{2}{\gamma_{00}} = -4\sum_{s=1}^{n} \overset{s}{\underset{2}{m}}(\overset{s}{r})^{-1},\tag{24}$$

where

$$\overset{s}{r}^2 = (x^k - \overset{s}{\xi^k})(x^k - \overset{s}{\xi^k})$$

is the distance (squared) of the space point (x^k) from the sth singularity $(\overset{s}{\xi^k})$, and the 'masses' $\overset{s}{\underset{2}{m}}$ can be functions of the time τ. That they are constants follows from the solvability requirement of the second of the equations (23)

$$\int \underset{2}{\gamma_{00,0k}}n_k\,\mathrm{d}S = 0.\tag{25}$$

Comparison with the Schwarzschild solution shows that they are in fact positive masses.

   This leads to a curious but only apparent difficulty whose resolution solves the problem of motion. According to the above, only monopole (with positive masses) solutions seem to be admitted but integrability of equations (20) (that is the validity of integrals (22)) requires addition of dipoles (and extra poles). The equations of motion, which are our aim, are obtained by removing these additional dipoles from the final result. The fictitious poles and dipoles can be added to a solution without affecting equation $(20a)$ because it has the form of a Poisson equation (with 'mass density' $2 \underset{2p-2}{N_{00}}$). Let now

$$\int_{2p-1}^{s} 2 \underset{2p-1}{N_{0k}} = 4\pi \underset{2p-1}{C_0} \quad \text{and} \quad \int_{2p}^{s} 2 \underset{2p}{N_{ik}} = 4\pi \underset{2p-1}{C_i} \tag{26}$$

where the C_μ are in general nonzero functions of τ (integrability of the equations (20) is assured if the C_μ vanish identically but this would make the system overdetermined).

By remembering that in the integrands of $\int^s$ only the terms 'r^{-2}' give in the limit a nonzero contribution, it is not difficult to find how the integrands in (26) should be changed in order that the newly obtained right hand sides (the functions $C_\mu \rightarrow C'_\mu$) vanish identically. Thus, having calculated $\underset{2p-2}{\gamma_{00}}$ from $(20a)$, we use the substitution

$$\underset{2p-2}{\gamma_{00}} \rightarrow \underset{2p-2}{\gamma_{00}} - \sum_s 4 \overset{s}{\underset{2p-2}{m}} (r)^{-1}, \tag{27}$$

where the $\overset{s}{\underset{2p-2}{m}}$ are multipliers (functions of τ) to be determined. Then

$$\overset{s}{\underset{2p-1}{C_0}} \rightarrow \overset{s}{\underset{2p-1}{C_0}} - 4 \overset{s}{\underset{2p-1}{\dot{m}}} \tag{28}$$

where the dot denotes differentiation with respect to τ. Similarly, if

$$\underset{2p-2}{\gamma_{00}} \rightarrow \underset{2p-2}{\gamma_{00}} - \sum_{s=1}^{n} \overset{s}{\underset{2p-2}{S_q}} (r)^{-1}_{,q} \tag{29}$$

(dipole replacement),

$$2\underset{2p}{N_{ij}} \rightarrow 2\underset{2p}{N_{ij}} + \sum_s \left[\overset{s}{\underset{2p}{\dot{S}_i}}(r)^{-1}_{,j} + \overset{s}{\underset{2p}{\dot{S}_j}}(r)^{-1}_{,i} \right.$$
$$\left. - \delta_{ij}\overset{s}{\underset{2p}{\dot{S}_q}}(r)^{-1}_{,q} \right] + \dots, \tag{30}$$

where terms omitted can be shown not to contribute to the surface integrals, and

$$\overset{s}{\underset{2p}{C_k}} \rightarrow \overset{s}{\underset{2p}{C_k}} - \overset{s}{\underset{2p}{\ddot{S}_k}}. \tag{31}$$

Hence, choosing

$$\overset{s}{\underset{2p-1}{\dot{m}}} = \overset{s}{\underset{2p-1}{C_0}} , \quad \overset{s}{\underset{2p}{\ddot{S}_k}} = \overset{s}{\underset{2p}{C_k}}, \tag{32}$$

we ensure that the integrals (22) vanish as required so that the equations (20) are integrable for all p.

Combining the two steps, we now see that

$$\gamma_{00} \rightarrow \gamma_{00} - \sum_p \lambda^{2p-2} \sum_{s=1}^{n} \left(4 \underset{2p-2}{\overset{s}{m}} (r)^{-1} + \underset{2p-2}{\overset{s}{S_q}} (r)^{-1}{}_{,q} \right). \tag{33}$$

The additional 'dipoles' can be removed therefore, if we take

$$\sum_p \lambda^{2p-2} \underset{2p-2}{\overset{s}{S_k}} = 0,$$

or

$$\sum_p \lambda^{2p} \underset{2p}{\overset{s}{\ddot{S}_k}} = \sum_p \lambda^{2p} \underset{2p}{\overset{s}{C_k}} = 0, \tag{34}$$

and the coefficient of $(\overset{s}{r})^{-1}$ in equation (33) given by

$$-\tfrac{1}{4}\left(\lambda^2 \underset{2}{\overset{s}{m}} + \lambda^4 \underset{4}{\overset{s}{m}} + \lambda^6 \underset{6}{\overset{s}{m}} + \dots \right) \tag{35}$$

is known because of the first of equations (32).

The sums (34) and (35) are necessarily finite, the number of terms terminating on the order to which an actual calculation is carried out. Indeed, equation (34) represents just the desired $3n$ equations of motion of a test particle.

We shall require in the sequel only the first approximation (called Newtonian and arising from λ^4) because the gravitational force between two particles, $m_1 m_2/r^2$, can be expected to be comparable to the Coulomb force between electrostatic charges. The cross term in the Lorentz force is not likely to appear in a quasi-stationary situation. In conclusion also, we may note that equations of motion can be obtained much more easily than by the EIH method, if we define 'point-mass' and 'dipole moment' in terms of integrals of $G_{\mu\nu}$ over a narrow tube surrounding a world line which the particle is supposed to follow. We can then simplify calculations by using a Fermi-coordinate system. In the absence of an energy–momentum tensor in the field equations, however, this method is difficult to motivate and has not been applied to GFT.

2. On Newtonian approximation and field strength

It may seem strange that any approximation procedure applied to the comprehensive field equations is capable of yielding equations of motion of a charged test particle. The strength of a Coulomb field is after all some 10^{39} times that of the gravitational field, for electrons at any rate. It is for this reason that the λ-expansion of EIH could not have anything to do with field strength. It was shown by Infeld and Wallace (in 1940) that it yields the correct, that is Lorentz equations of motion, when used on the general relativistic, Einstein–Maxwell field equations. The argument of this monograph is, however, that this is not enough to establish the validity of Einstein–Maxwell theory as a description of electromagnetism is curved space–time. Derivability of equations of motion from the nonlinear field equations may be a necessary validity condition but it is not necessary and sufficient. The result of Infeld and Wallace also fails as a test of the theory because it predicts only what is well known from nonrelativistic physics.

When I said in the last section that the Newtonian and Coulomb forces are comparable, I was referring to the order in λ in which they could be expected to appear and not to comparative field strength. But what does the statement that electromagnetic interactions are stronger (by a factor of 10^{39}) than gravitational mean? There is an argument which shows that it is a statement about the nature of electrons rather than about an absolute magnitude of forces to which a given field gives rise. If we calculate the electrostatic energy of two electrons in contact (assuming a finite radius of a spherical model of the particle, say about 10^{-13} cm) and convert it into mass by the special relativistic formula, the result is of the same order of magnitude as the electronic mass. This may be a more realistic comparison than a direct calculation because energy is the only thing which mass and charge have in common.

Let us return now to the gravitational case and consider two particles of (gravitational) mass $\overset{1}{m}, \overset{2}{m}$ respectively. We can write then

$$\overset{1}{r}{}^2 = (x^k - \eta^k)^2, \quad \overset{2}{r}{}^2 = (x^k - \zeta^k)^2. \tag{1}$$

The lowest order of approximation in λ is then given by

$$p = 2. \tag{2}$$

The equations (1.20) become

$$\left. \begin{aligned} &\underset{2}{\gamma}_{00,jj} = 0, \quad \underset{3}{\gamma}_{0k,jj} = 0, \\ &\underset{4}{\gamma}_{ij,kk} = -\underset{3}{\gamma}_{0i,0j} - \underset{3}{\gamma}_{0j,0i} + 2\delta_{ij}\varphi_{,00} - 2\varphi\varphi_{,ij} \\ &\qquad\qquad - \varphi_{,i}\varphi_{,j} + \tfrac{3}{2}\delta_{ij}\varphi_{,k}\varphi_{,k},) \end{aligned} \right\} \tag{3}$$

where

$$\varphi = -2\left(\frac{\overset{1}{m}}{\overset{1}{r}} + \frac{\overset{2}{m}}{\overset{2}{r}}\right). \tag{4}$$

We impose coordinate conditions (as we may because of Bianchi identities)

$$\underset{3}{\gamma_{0k,k}} - \underset{2}{\gamma_{00,0}} = 0 = \underset{4}{\gamma_{kj,j}}, \tag{5}$$

and it is then not sufficient to show that the equations of motion

$$\underset{4}{\overset{1}{C'}_k} = 0 = \underset{4}{\overset{2}{C}_k} \tag{6}$$

are

$$\ddot{\eta}^k + \frac{\partial}{\partial \eta^k}\frac{\overset{2}{m}}{r} = 0 = \ddot{\zeta}^k + \frac{\partial}{\partial \zeta^k}\frac{\overset{1}{m}}{r}, \tag{7}$$

where

$$r^2 = (\eta^k - \zeta^k)^2,$$

and, because of (1.25), $\overset{1}{m}$ and $\overset{2}{m}$ are necessarily constant.

3.   Equations of motion in GFT

The method of Einstein, Infeld and Hoffman was extended by Infeld (1950) to the field equations of the nonsymmetric theory. Actually, he considered only the strong field equations which result from WFE (2.3.17) if

$$\Gamma_\mu = \varphi_{,\mu}, \tag{1}$$

that is Γ_μ is the gradient of a scalar function of position. The weak field equations were tackled by Callaway (1953). Let us write as before

$$g_{(\mu\nu)} = h_{\mu\nu}, \quad g_{[\mu\nu]} = k_{\mu\nu},$$

and assume with Infeld, that

$$\left.\begin{aligned}
h_{00} &= 1 + \lambda^2 \underset{2}{h_{00}} + \lambda^4 \underset{4}{h_{00}} + \cdots, \\
h_{ij} &= -\delta_{ij} + \lambda^2 \underset{2}{h_{ij}} + \lambda^4 \underset{4}{h_{ij}} + \cdots, \\
h_{0k} &= \lambda^3 \underset{3}{h_{0k}} + \lambda^5 \underset{5}{h_{0k}} + \cdots, \quad k_{0j} = \lambda^3 \underset{3}{k_{0j}} + \lambda^5 \underset{5}{k_{0j}} + \cdots, \\
k_{ij} &= \lambda^2 \underset{2}{k_{ij}} + \lambda^4 \underset{4}{k_{ij}} + \cdots.
\end{aligned}\right\} \tag{2}$$

Let us also write

$$\tilde{\Gamma}^{\lambda}_{\mu\nu} = \left\{ \begin{array}{c} \lambda \\ \mu\nu \end{array} \right\}_h + M^{\lambda}_{\mu\nu}. \tag{3}$$

I must point out that this is a purely formal separation which does not prejudice our choice of the metric in chapter 3. The tensor $h_{\mu\nu}$, albeit nonsingular, is not the metric, whatever we may do with it. This is contrary to what Infeld asserted and to what was repeated in the original article by Russell and myself (1972b) on the equations of motion.

If we reserve a semicolon without subscripts for a covariant derivative with respect to $\left\{ \begin{array}{c} \lambda \\ \mu\nu \end{array} \right\}_h$, then

$$h_{\mu\nu;\lambda} = 0.$$

Then, from

$$g_{\mu\nu;\lambda}(\tilde{\Gamma}) = 0,$$
$$\phantom{g_{\mu\nu;\lambda}}{}_{+\,-}$$

we find that

$$k_{\mu\nu;\lambda} - M^{\sigma}_{\mu\lambda}(h_{\sigma\nu} + k_{\sigma\nu}) - M^{\sigma}_{\lambda\nu}(h_{\mu\sigma} + k_{\mu\sigma}) = 0, \tag{4}$$

whence, in the usual way,

$$M^{\lambda}_{\mu\nu} = h^{\lambda\sigma}[k_{\mu\nu;\sigma} - I_{\mu\nu\sigma} - M^{\rho}_{\nu\sigma}k_{\mu\rho} - M^{\rho}_{\sigma\mu}k_{\rho\nu}], \tag{5}$$

where

$$I_{\mu\nu\sigma} = \tfrac{1}{2}k_{[\mu\nu;\sigma]} = \tfrac{1}{2}k_{[\mu\nu,\sigma]}. \tag{6}$$

Because of the expansion (2), it follows immediately that

$$\left. \begin{array}{l} \underset{2}{M^{0}}_{00} = \underset{2}{M^{0}}_{0k} = \underset{2}{M^{0}}_{k0} = \underset{3}{M^{0}}_{00} = 0, \\[2mm] \underset{3}{M^{i}}_{0j} = -\underset{3}{k}_{0j,i} + \underset{3}{I}_{0ji}, \quad \underset{3}{M^{0}}_{ij} = \underset{3}{k}_{ij,0} - \underset{3}{I}_{ij0}, \\[2mm] \underset{2}{M^{i}}_{jk} = -\underset{2}{k}_{jk,i} + \underset{2}{I}_{jki}, \end{array} \right\} \tag{7}$$

and therefore

$$\left. \begin{array}{l} \underset{4}{M^{k}}_{00} = 0, \quad \underset{4}{M^{0}}_{0k} = \underset{3}{k}_{0k,0}, \\[3mm] \underset{4}{M^{i}}_{jk} = -\underset{4}{k}_{jk,i} + \underset{4}{I}_{jk,i} + \underset{2}{h}^{ip}\left(\underset{2}{k}_{jk,p} - \underset{2}{I}_{jkp}\right) - \underset{2}{k}_{js}\left\{\begin{array}{c} s \\ ik \end{array}\right\}_{2} - \underset{2}{k}_{sk}\left\{\begin{array}{c} s \\ ji \end{array}\right\}_{2} \\[4mm] \qquad - \left[\underset{2}{k}_{js}\left(-\underset{2}{k}_{ki,s} + \underset{2}{I}_{kis}\right) + \underset{2}{k}_{sk}\left(-\underset{2}{k}_{ij,s} + \underset{2}{I}_{ijs}\right)\right]. \end{array} \right\} \tag{8}$$

The fields represented by $h_{\mu\nu}$ and $k_{\mu\nu}$ decouple in λ^2 and, as we have seen, it is necessary to go to λ^4 in order to obtain the 'gravitational' term $\overset{1}{m}\overset{2}{m}/r^2$. In view of this, we must now consider the 'electromagnetic' correction of the equations of motion arising from the fourth power of the parameter λ.

Let $P_{\mu\nu}$ denote the symmetric Ricci tensor formed from the Christoffel brackets $\left\{ \begin{matrix} \lambda \\ \mu\nu \end{matrix} \right\}_h$. Using equations (8) it can be shown directly that

$$\underset{4}{R_{00}} - \underset{4}{P_{00}} = 0, \tag{9}$$

and that

$$\underset{4}{R_{(ij)}} - \underset{4}{P_{ij}} = \left[\underset{}{k_{is}} \left(\underset{2}{I_{jps}} - \underset{2}{k_{jp,s}} \right) + \underset{}{k_{sj}} \left(\underset{2}{I_{pis}} - \underset{2}{k_{pi,s}} \right) \right]_{,p}$$

$$- \tfrac{1}{2} \left\{ \left[\underset{}{k_{sp}} \left(\underset{2}{I_{pis}} - \underset{2}{k_{pi,s}} \right) \right]_{,j} + \left[\underset{}{k_{sp}} \left(\underset{2}{I_{pjs}} - \underset{2}{k_{pj,s}} \right) \right]_{,i} \right\}$$

$$- \left(\underset{2}{I_{ips}} - \underset{2}{k_{ip,s}} \right) \left(\underset{2}{I_{sjp}} - \underset{2}{k_{sj,p}} \right). \tag{10}$$

But, from the general theory of section 1, it follows readily that the equations of motion are given by

$$0 = - 2\pi \overset{s}{\underset{4}{c_i}} = \int^s N_{ij} n^j \mathrm{d}S. \tag{11}$$

The integral (11) is taken as before over a 2-sphere surrounding the sth source of the field and

$$N_{ij} = \underset{4}{R_{(ij)}} - \underset{4}{P_{ij}} + \tfrac{1}{2}\delta_{ij} \left(\underset{4}{R_{00}} - \underset{4}{P_{00}} \right) - \tfrac{1}{2}\delta_{ij} \left(\underset{4}{R_{ss}} - \underset{4}{P_{ss}} \right)$$

$$= \underset{4}{R_{(ij)}} - \underset{4}{P_{ij}} - \tfrac{1}{2}\delta_{ij} \left(\underset{4}{R_{ss}} - \underset{4}{P_{ss}} \right), \tag{12}$$

because of (9).

Let us now consider the problem of two particles carrying masses $\overset{1}{m}, \overset{2}{m}$ and charges $\overset{1}{e}, \overset{2}{e}$ respectively. Let us also assume (as in section 2) that for either mass and either charge,

$$m = \lambda^2 \underset{2}{m} + \lambda^4 \underset{4}{m} + \ldots, \tag{13}$$

and

$$e = \lambda^2 \underset{2}{e} + \lambda^4 \underset{4}{e} + \dots \tag{14}$$

Let also $\underset{2}{\overset{1}{f}}_{\mu\nu}$ and $\underset{2}{\overset{2}{f}}_{\mu\nu}$ be the electromagnetic fields due to the first and to the second source respectively. Then

$$\underset{2}{f}_{\mu\nu} = \underset{2}{\overset{1}{f}}_{\mu\nu} + \underset{2}{\overset{2}{f}}_{\mu\nu}. \tag{15}$$

Classical Maxwell theory requires that when the sources are quasi-stationary, as implied by the λ method anyway, the electrostatic potential for from either source is given approximately by

$$\varphi \propto \left(\underset{2}{\overset{1}{e}} + \underset{2}{\overset{2}{e}} \right) \Big/ r \tag{16}$$

and that

$$\underset{2}{f}_{\mu\nu} \text{ is proportional to } \varepsilon_{\mu\nu\lambda}\varphi_{,\lambda} = \varepsilon_{\mu\nu\kappa}\varphi_{,\kappa}. \tag{17}$$

Now, up to λ^2, the weak field equations give

$$\underset{2}{k}_{is,s} = 0, \tag{18}$$

and

$$\underset{2}{k}_{[ij,k],ss} = 0. \tag{19}$$

This suggests that, at least to the second order in λ, the electromagnetic field tensor should be identified as

$$\underset{2}{f}_{ij} = A \underset{2}{k}_{ij,ss}, \tag{20}$$

where A is a constant which can be set equal to unity by a free choice of units. We can write then

$$\underset{2}{k}_{ij,ss} = \varepsilon_{ijk}\varphi_{,k}. \tag{21}$$

Let us now define a function Φ by the quasi-Poisson equation

$$\Phi_{,ss} = \varphi, \tag{22}$$

so that

$$\underset{2}{k}_{ij} = \varepsilon_{ijk}\Phi_{,k}. \tag{23}$$

But φ is by definition (or rather, choice) a harmonic function and therefore the weak field equations are satisfied in λ^2. Also, the only component of I_{ijk} is

$$I_{123} = \tfrac{3}{2}\varphi, \tag{24}$$

and, from equations (12), (10) and (9)

$$N'_{ij} = F_{ijp,p} - S_{ij},$$
$$\underset{4}{} \tag{25}$$

where

$$F_{ijp} = -\varepsilon_{isa}\varepsilon_{jpb}\Phi_{,a}\Phi_{,sb} - \delta_{pi}\Phi_{,aj}\Phi_{,a}$$
$$+ \delta_{ij}\Phi_{,ap}\Phi_{,a} + \tfrac{1}{2}\delta_{ip}\varphi_{,j}\Phi - \tfrac{1}{2}\delta_{ij}\varphi_{,p}\Phi, \tag{26}$$

and

$$S_{ij} = \tfrac{1}{4}\delta_{ij}\varphi^2 - \tfrac{1}{2}\varphi_{,i}\Phi_{,j} + \tfrac{1}{2}\varphi_{,ij}\Phi, \tag{27}$$

and a dash indicates the electromagnetic correction to equations (2.7).

We can see two things at once. First

$$F_{ijp} = -F_{ipj}. \tag{28}$$

This is the only term obtained by Infeld, and its skewsymmetry guaranteed that it cancelled out from the 2-integral (11) leading to a null result as far as the Lorentz force is concerned. The new term

$$\frac{1}{2\pi} \int^s S_{ij}n^j \mathrm{d}S, \tag{29}$$

however, does not vanish, or at least we can make sure that it does not. It can be verified at once that it is independent of the shape of the surface of the source because

$$S_{ij,j} = 0. \tag{30}$$

We shall derive the explicit form of the Lorentz force in the next section.

4.   A new law of motion

Let us now drop the small numbers indicating that only the first terms from the expansions (3.13) and (3.14) for mass and charge appear in the Newtonian approximation, and make them subscripts. Thus, we have two particles with charges e_1 and e_2, and masses m_1 and m_2 respectively. Let them be located at a time τ at the points whose space coordinates are ξ^i and η^i respectively. If x^i are the space coordinates of a field point, let

$$x^i - \xi^i = R^i, \quad x^i - \eta^i = \rho^i. \tag{1}$$

For quasi-stationary point particles, the general solution for the biharmonic function Φ has the form $(i = 1, 2)$

$$\Phi_i = c_1 e_i r^2 + c_2 e_i r + c_3 e_i + c_4 e_i/r \tag{2}$$

where

$$r = R = \sqrt{(R^i R^i)} \text{ or } \rho = \sqrt{(\rho^i \rho^i)} \tag{3}$$

and c_1, c_2, c_3, c_4 are constants of integration.

Consider the motion of the first particle (m_1, e_1, R^i). We surround it by a sphere of radius R and retain in S_{ij} only those terms which are of the order R^{-2} since, as shown by Einstein & Infeld, they alone contribute to the surface integrals. We have

$$\Phi = \Phi_1 + \Phi_2 = c_1 e_1 R^2 + c_1 e_2 \rho^2 + c_2 e_1 R + c_2 e_2 \rho + c_3 e_1$$
$$+ c_3 e_2 + c_4 e_1 R^{-1} + c_4 e_2 \rho^{-1}, \tag{4}$$

and the only surviving terms in S_{ij} are

$$2 c_1 c_2 e_1 e_2 \rho^j R^k R^{-3} + c_2^2 e_1 e_2 \rho^j R^k \rho^{-1} R^{-3}$$
$$- c_2 c_4 e_1 e_2 \rho^j R^k \rho^{-3} R^{-3}$$
$$- c_2 c_4 e_1 e_2 \rho^k R^j \rho^{-3} R^{-3} + 4 c_2^2 e_1^2 R^j R^k R^{-4}. \tag{5}$$

Hence, integrating over the sphere (and adding to the corresponding integral obtained in GR, in sections 1 and 2), the equations of motion of the first particle become

$$m_1 \ddot{\zeta}^j + m_1 m_2 \rho^j \rho^{-3} = \tfrac{1}{3} c_1 c_2 e_1 e_2 \rho^j + \tfrac{1}{6} c_2^2 e_1 e_2 \rho^j \rho^{-1}$$
$$- \tfrac{2}{3} c_2 c_4 e_1 e_2 \rho^j \rho^{-3}, \tag{6}$$

or, if r denotes its position vector relative to an origin which coincides instantaneously with the second particle,

$$m_1 \frac{\mathrm{d}^2 r}{\mathrm{d}\tau^2} + \frac{m_1 m_2 \hat{r}}{r^2} = \tfrac{1}{3} e_1 e_2 (c_1 c_2 r + \tfrac{1}{2} c_2^2) \hat{r}$$
$$- \tfrac{2}{3} e_1 e_2 c_2 c_4 \frac{\hat{r}}{r^2}, \tag{7}$$

where $\hat{r}$ is a unit vector in the r direction. The equation of motion of the second particle in the field of the first is, of course, obtained by interchanging ζ with η, and m_1 with m_2. It now follows that the Lorentz force will not vanish identically provided

$$c_2 \neq 0, \tag{8}$$

and for the Coulomb force to be present, it is also necessary that

$$c_4 \neq 0. \tag{9}$$

The integration constants are therefore not altogether arbitrary. Furthermore, there is no reason why c_1 should vanish as was required by Treder, who in 1957 first considered a biharmonic potential, on the grounds that if $c_1 \neq 0$ the field would increase with r. So far, our work differs little from Treder's though I shall show in the next section that its physical basis is entirely new.

Let us dispose first of the problem created by a nonzero c_1. If we require that charges be measured in electrostatic units, we should have

$$e_2 c_4 = -\tfrac{3}{2}. \tag{10}$$

Let us further suppose that as r tends to r_0, which is large but finite to avoid Treder's difficulty, k_{23} (and therefore also k_{23}) tends to zero. Then, from (2) and (3.23), we can expect that

$$2c_1 c_4 r_0 - \tfrac{2}{3} - e_4^2/r_0^2 = 0. \tag{11}$$

Hence c_1 can not be zero under any circumstances. We can define new constants, c_5 and c_6, by

$$c_1 = c_5 r_0^{-2} \quad \text{and} \quad c_4 = -3/2c_2 = c_6 r_0. \tag{12}$$

If we also write

$$c_6 = 2\mu c_5, \tag{13}$$

then from the condition (11)

$$c_5^2 = 3/8\mu(1 - \mu), \quad 0 < \mu < 1, \tag{14}$$

and the magnitude of the Lorentz force becomes finally

$$e_1 e_2 \left[\frac{1}{r^2} + \frac{r_0 \lambda - r}{4(1 - \lambda)r_0^3} \right], \tag{15}$$

where

$$\lambda = 1 - \mu.$$

Since we can expect the correction $(r_0\lambda - r)/4(1 - \lambda)r_0^3$ to the Coulomb force to be small, we can tentatively identify r_0 as the radius of a finite universe. We are anticipating subsequent cosmological developments of GFT but note that an infinite r_0 creates more problems than it solves, even now.

5.   Conditions for an identification of
the electromagnetic field tensor

Somewhat surprisingly, solution of the problem of motion enables us to identify the electromagnetic field tensor within the structure of the generalised field theory. I have said that the physical basis of the foregoing integration of the field equations is quite different from the otherwise very similar work of Treder. The reason for this is that Treder postulated the biharmonic expression for electrostatic potential virtually without justification. On the other hand, the weak principle of geometrisation together with the metric hypothesis of chapter 3 imply that since $g_{(\mu\nu)}$ is not the metric tensor, there is no compulsion to identify $g_{[\mu\nu]}$ as an electromagnetic field tensor either. (It can be easily verified that if this choice is made then S_{ij} does not contribute to the integral around the particle and no Lorentz force arises in the equations of motion of the order λ^4.)

Let us consider, instead of the λ expansion of EIH, the more general expansions

$$g_{\mu\nu} = \eta_{\mu\nu} + \varepsilon h_{\mu\nu} + \varepsilon^2 q_{\mu\nu} + \varepsilon^3 \alpha_{\mu\nu} + \ldots, \tag{1}$$

and

$$g^{\mu\nu} = \eta^{\mu\nu} + \varepsilon H^{\mu\nu} + \varepsilon^2 Q^{\mu\nu} + \varepsilon^3 A^{\mu\nu} + \ldots, \tag{2}$$

where $\eta_{\mu\nu}$ is as always the Minkowski metric tensor, but no restriction is put on the other quantities. It is not necessary to regard the above as a weak field approximation as long as we allow comparison of different orders of magnitude in the otherwise free parameter ε. Let all tensor indices be raised and lowered with the Minkowski tensor, remembering that this is only a notational device which does not prejudice the choice of a metric. Thus we write

$$h^{\mu}{}_{\nu} \overset{\text{Df}}{=} \eta^{\mu\sigma} h_{\sigma\nu}, \quad \alpha^{\mu\nu} \overset{\text{Df}}{=} \eta^{\mu\rho} \eta^{\nu\sigma} \alpha_{\rho\sigma} \text{ etc.} \tag{3}$$

The conditions for a tensor $f_{\mu\nu}$ to be identifiable as a Maxwell tensor which arise from the problem of motion were formulated in 1977 by L. J. Gregory and myself (1977b). They are

$$f_{\mu\nu} = -f_{\nu\mu}, \tag{4a}$$

which is self-evident,

$$\underset{2}{f}_{\mu\nu} = \Box q_{[\mu\nu]} \tag{4b}$$

in the order ε^2, where $\Box$ is the D'Alembertian operator, and

$$f_{\mu\nu} \text{ contains second derivatives of } g_{[\mu\nu]}. \tag{4c}$$

The point of the third condition is that $f_{\mu\nu}$ should be identified as an exact quantity in terms of the fields $g_{\mu\nu}$, Γ_{μ} and the connection $\tilde{\Gamma}^{\lambda}{}_{\mu\nu}$. This distinguishes GFT yet again from the work of Treder.

There are three expressions which can be considered. Historically, Russell and I let $f_{\mu\nu}$ be proportional to the tensor

$$\tilde{w}_{\mu\nu} = g^{\alpha\beta}{}_{[\mu\nu];\alpha\beta} \tag{5}$$

but $\tilde{w}_{\mu\nu}$ is not skewsymmetric in general. All of the conditions (4) are, however, satisfied by the tensor

$$w_{\mu\nu} = g^{\alpha\beta} g_{[\mu\nu];\alpha\beta}. \tag{6}$$

It can be easily verified also that

$$\begin{aligned} R_{[\mu\nu]}(\tilde{\Gamma}) &= \tfrac{2}{3}(\Gamma_{\mu,\nu} - \Gamma_{\nu,\mu}) \\ &= -\tilde{\Gamma}^{\sigma}{}_{[\mu\nu];\sigma} \\ &= C^{\alpha\beta} g_{[\mu\nu],\alpha\beta} + \ldots, \end{aligned} \tag{7}$$

where $C^{\alpha\beta}$ is a symmetric tensor whose exact form is immaterial to our problem. It follows from the last of equations (7) that $R_{[\mu\nu]}$ satisfies the third of the above conditions. But from the expansion (1), it follows that

$$R_{2[\mu\nu]} = \tfrac{1}{2}\Box q_{[\mu\nu]} \tag{8}$$

so that the second condition is also satisfied by this skew tensor. Hence we can have

$$f_{\mu\nu} \propto R_{[\mu\nu]}(\tilde{\Gamma}), \tag{9}$$

as well. The choice between the tensors $w_{\mu\nu}$ and $R_{[\mu\nu]}$ must now be made on the grounds of consequences implied by particular solutions of the field equations. One thing, however, stands out immediately and that is that $R_{[\mu\nu]}$ is exactly the curl of a 4-vector ($\tfrac{2}{3}\Gamma_\mu$). A more detailed comparison between the two tensors will be made in the next section in terms of the weak field expansion (1) to see the extent to which $w_{\mu\nu}$ can be approximated by a curl. The tensor $w_{\mu\nu}$ is not a very good candidate for the electromagnetic intensity for yet another reason. The well-known Papapetrou solution (1948),

$$-g_{(11)}^{(-1)} = g_{(00)} = 1 - 2m/r, \quad g_{(22)} = -r^2,$$

$$g_{[23]} = -cr^2; \quad c, m \text{ constants}, \tag{10}$$

gives, for the radial electrostatic field, a pseudo-Coulomb force of magnitude proportional to

$$\frac{1}{r^2}\left(1 - \frac{2m}{r}\right), \tag{11}$$

which is difficult to reconcile with the Lorentz force derived previously:

$$(1/r^2) + k(r_0 - r). \tag{12}$$

6. The tensors $w_{\mu\nu}$ and $R_{[\mu\nu]}(\tilde{\Gamma})$

The relation

$$g^{\mu\sigma}g_{\nu\sigma} = \delta^\mu_{\ \nu} = g^{\sigma\mu}g_{\sigma\nu}$$

gives by inserting into it the expansions (5.1) and (5.2),

$$H^{\rho\sigma} = -h^{\sigma\rho}, \quad Q^{\rho\sigma} = -q^{\sigma\rho} + h^{\sigma\alpha}h_\alpha^{\ \rho} \tag{1}$$

and

$$A^{\rho\sigma} = -\alpha^{\sigma\rho} - h_\beta^{\ \rho}h^{\sigma\gamma}h_\gamma^{\ \beta} + h_{\beta\alpha}(\eta^{\sigma\alpha}q^{\beta\rho} + \eta^{\rho\alpha}q^{\alpha\beta}).$$

It is convenient, though not necessary, to assume that

$$h_{\mu\nu} = h_{(\mu\nu)}, \tag{2}$$

when also

$$Q^{[\mu\nu]} = q^{[\mu\nu]}. \tag{3}$$

(We remember that tensor indices are raised and lowered with $\eta_{\mu\nu}$.) A straightforward calculation gives

$$w_{\mu\nu} = g^{\alpha\beta}[\tilde{\Gamma}^{\sigma}_{[\mu\alpha];\beta}g_{(\sigma\nu)} + \tilde{\Gamma}^{\sigma}_{[\alpha\nu];\beta}g_{(\mu\sigma)} + \tilde{\Gamma}^{\sigma}_{[\mu\alpha]}\tilde{\Gamma}^{\rho}_{[\sigma\beta]}g_{[\rho\nu]}$$
$$+ \tilde{\Gamma}^{\sigma}_{[\alpha\nu]}\tilde{\Gamma}^{\rho}_{[\beta\sigma]}g_{[\mu\rho]}] + g^{(\alpha\beta)}g_{[\sigma\rho]}(\tilde{\Gamma}^{\sigma}_{[\mu\alpha]}\tilde{\Gamma}^{\rho}_{[\beta\nu]}$$
$$- \tilde{\Gamma}^{\sigma}_{[\alpha\nu]}\tilde{\Gamma}^{\rho}_{[\mu\beta]}), \tag{4}$$

where semicolons denote covariant differentiation with respect to $\tilde{\Gamma}^{\lambda}_{(\mu\nu)}$. Let us now assume that the expansion (5.1) induces

$$\tilde{\Gamma}^{\lambda}_{\mu\nu} = \underset{0}{\tilde{\Gamma}}^{\lambda}_{\mu\nu} + \varepsilon \underset{1}{\tilde{\Gamma}}^{\lambda}_{\mu\nu} + \varepsilon^2 \underset{2}{\tilde{\Gamma}}^{\lambda}_{\mu\nu} + \varepsilon^3 \underset{3}{\tilde{\Gamma}}^{\lambda}_{\mu\nu} + \dots \tag{5}$$

Then, equating to zero coefficients of successive powers of ε in

$$g_{\mu\nu;\lambda}(\tilde{\Gamma}) = 0,$$
$$\underset{+ \; -}{}$$

and solving the resulting equations, we get

$$\underset{0}{\tilde{\Gamma}}^{\lambda}_{\mu\nu} = 0,$$
$$\left. \begin{array}{l} \underset{1}{\tilde{\Gamma}}^{\lambda}_{\mu\nu} = \underset{1}{\tilde{\Gamma}}^{\lambda}_{\nu\mu} = \tfrac{1}{2}\eta^{\lambda\sigma}(h_{\sigma\nu,\mu} + h_{\mu\sigma,\nu} - h_{\mu\nu,\sigma}), \\[2mm] \underset{2}{\tilde{\Gamma}}^{\lambda}_{(\mu\nu)} = \tfrac{1}{2}\eta^{\lambda\sigma}(m_{\sigma\mu,\nu} + m_{\nu\sigma,\mu} - m_{\mu\nu,\sigma}), \\[2mm] \underset{2}{\tilde{\Gamma}}^{\lambda}_{[\mu\nu]} = \tfrac{1}{2}\eta^{\lambda\sigma}(q_{\mu\nu,\sigma} + q_{\sigma\nu,\mu} + q_{\mu\sigma,\nu}), \\[2mm] \underset{3}{\tilde{\Gamma}}^{\lambda}_{(\mu\nu)} = \tfrac{1}{2}\eta^{\lambda\sigma}(n_{\mu\nu\sigma} + n_{\sigma\nu\mu} - n_{\nu\mu\sigma}), \end{array} \right\} \tag{6}$$

and

$$\underset{3}{\tilde{\Gamma}}^{\lambda}_{[\mu,\nu]} = \eta^{\lambda\sigma}[\tfrac{1}{2}(\alpha_{[\mu\sigma],\nu} + \alpha_{[\sigma\nu],\mu} + \alpha_{[\mu\nu],\lambda})$$
$$- \underset{1}{\tilde{\Gamma}}^{\rho}_{\mu\sigma}q_{[\rho\nu]} - \underset{1}{\tilde{\Gamma}}^{\rho}_{[\sigma\nu]}q_{[\mu\rho]} - \underset{2}{\tilde{\Gamma}}^{\rho}_{[\mu\nu]}h_{\rho\lambda}]$$

where

$$m_{\mu\nu} = q_{(\mu\nu)} - h_{\mu\nu},$$
$$n_{\lambda\mu\nu} = \alpha_{\lambda\mu,\nu} - \underset{1}{\tilde{\Gamma}}^{\sigma}_{\lambda\nu}q_{\sigma\mu} - \underset{1}{\tilde{\Gamma}}^{\sigma}_{\nu\mu}q_{\lambda\sigma} - \underset{2}{\tilde{\Gamma}}^{\sigma}_{\lambda\nu}h_{\sigma\mu} - \underset{2}{\tilde{\Gamma}}^{\sigma}_{\nu\mu}h_{\lambda\sigma}.$$

We can verify easily that

$$Q^{[\mu\nu]},_{\nu} = -\eta^{\nu\rho}\eta^{\mu\sigma}q_{[\rho\sigma],\nu}$$
$$= \eta^{\nu\rho}\eta^{\mu\sigma}(\underset{2}{\tilde{\Gamma}}^{\alpha}_{[\rho\nu]}\eta_{\alpha\sigma} + \underset{2}{\tilde{\Gamma}}^{\alpha}_{[\nu\sigma]}\eta_{\rho\alpha})$$
$$= -\eta^{\mu\sigma}\underset{2}{\tilde{\Gamma}}_{\sigma} = 0, \tag{7}$$

as we know in any case since $g^{[\mu\nu]},_{\nu} = 0 = \tilde{\Gamma}_{\mu}$. The expansion of the equation

$$g^{[\mu\nu]},_{\nu} = 0,$$

therefore, starts with ε^3, and equating to zero its coefficient gives

$$(A^{[\mu\nu]} + hQ^{[\mu\nu]})_{,\nu} = 0. \tag{8}$$

Another useful relation can be obtained by defining

$$D^{\sigma}{}_{\mu\nu} = \eta^{\sigma\beta}\eta_{\alpha\nu}\underset{1}{\tilde{\Gamma}}{}^{\alpha}{}_{\beta\mu} - \underset{1}{\tilde{\Gamma}}{}^{\sigma}{}_{\mu\nu}, \tag{9}$$

and

$$D_{\lambda\mu\nu} = \eta_{\lambda\sigma}D^{\sigma}{}_{\mu\nu} = h_{\mu\nu,\lambda} - h_{\lambda\mu,\nu}, \tag{10}$$

when

$$(\eta^{\nu\tau}\eta^{\rho\lambda} + \eta^{\nu\rho}\eta^{\tau\lambda})D_{\lambda\mu\nu} = 0. \tag{11}$$

The geometrical connection $\Gamma^{\lambda}{}_{\mu\nu}$ can not be expressed in terms of $\tilde{\Gamma}^{\lambda}{}_{\mu\nu}$. We must therefore assume that its expansion is of the same form as that of the Schrödinger connection. However, we can write

$$w_{\mu\nu} = \varepsilon^2\underset{2}{w}_{\mu\nu} + \varepsilon^3\underset{3}{w}_{\mu\nu} + \dots \tag{12}$$

Then, we easily find that

$$\begin{aligned}
\underset{2}{w}_{\mu\nu} &= \eta^{\alpha\beta}\left[\underset{2}{\tilde{\Gamma}}{}^{\sigma}{}_{[\mu\alpha],\beta}\eta_{\sigma\nu} + \underset{2}{\tilde{\Gamma}}{}^{\sigma}{}_{[\alpha\nu],\beta}\eta_{\mu\sigma} \right] \\
&= 2\underset{2}{\tilde{\Gamma}}{}^{\sigma}{}_{[\mu\nu],\sigma} \\
&= 2\underset{2}{\tilde{\Gamma}}{}^{\sigma}{}_{[\mu\nu];\sigma},
\end{aligned} \tag{13}$$

since

$$\underset{2}{\tilde{\Gamma}}{}^{\sigma}{}_{[\mu\alpha],\beta}\eta_{\sigma\nu}\eta^{\alpha\beta} = \tfrac{1}{2}\eta^{\alpha\beta}q_{[\mu\nu],\alpha\beta} = \underset{2}{\tilde{\Gamma}}{}^{\sigma}{}_{[\mu\nu],\sigma}. \tag{14}$$

Hence

$$\underset{2}{w}_{\mu\nu} = \tfrac{4}{3}\left[\underset{2}{\Gamma}_{\nu,\mu} - \underset{2}{\Gamma}_{\mu,\nu} \right]. \tag{15}$$

Also, after a straightforward, though somewhat more complicated, calculation

$$\underset{3}{w}_{\mu\nu} = \varphi_{\nu,\mu} - \varphi_{\mu,\nu} + \Omega_{\mu\nu}, \tag{16}$$

where

$$\varphi_{\mu} = \tfrac{4}{3}\underset{3}{\Gamma}_{\mu} + h^{\alpha\beta}q_{[\alpha\mu],\beta} - \eta^{\alpha\beta}\underset{1}{\tilde{\Gamma}}{}^{\sigma}{}_{\alpha\beta}q_{[\sigma\mu]}, \tag{17}$$

and

$$\begin{aligned}
\Omega_{\mu\nu} = \eta^{\alpha\beta}\Big[&3q_{[\alpha\sigma]}\underset{1}{\tilde{\Gamma}}{}^{\sigma}{}_{\mu\beta,\nu} - 3q_{[\alpha\sigma]}\underset{1}{\tilde{\Gamma}}{}^{\sigma}{}_{\nu\beta,\mu} \\
&+ q_{[\sigma\nu]}\underset{1}{\tilde{\Gamma}}{}^{\sigma}{}_{\mu\alpha,\beta} + q_{[\alpha\nu]}\underset{1}{\tilde{\Gamma}}{}^{\sigma}{}_{\sigma\beta,\mu} + q_{[\sigma\nu]}\underset{1}{\tilde{\Gamma}}{}^{\sigma}{}_{\alpha\beta,\mu} \\
&- q_{[\sigma\mu]}\underset{1}{\tilde{\Gamma}}{}^{\sigma}{}_{\nu\alpha,\beta} - q_{[\alpha\mu]}\underset{1}{\tilde{\Gamma}}{}^{\sigma}{}_{\sigma\beta,\nu} - q_{[\sigma\mu]}\underset{1}{\tilde{\Gamma}}{}^{\sigma}{}_{\alpha\beta,\nu} \Big].
\end{aligned} \tag{18}$$

Hence, although

$$w_{\mu\nu} = \tfrac{4}{3}\left(\Gamma_{\nu,\mu} - \Gamma_{\mu,\nu}\right) = 2R_{[\nu\mu]}(\tilde{\Gamma}), \tag{19}$$

$w_{\mu\nu}$ is not, in general, the curl of a vector.

7.   A non-Maxwellian electrodynamics

I am now going to reject the identification of $w_{\mu\nu}$ as the electromagnetic intensity tensor and propose that

$$f_{\mu\nu} = kR_{[\mu\nu]}(\tilde{\Gamma}), \tag{1}$$

where k is a constant which can not be readily determined in the generalised field theory without a further assumption. Then Γ_μ becomes proportional to the electromagnetic 4-vector potential and one set of Maxwell's equations (usually called the second)

$$f_{[\mu\nu,\lambda]} = 0 \tag{2}$$

is identically satisfied. However, do we have Maxwell's electro-dynamics in full? The answer depends on the identification of a ('Maxwellian') material tensor and on the definition of a current vector density. There are certain reasons (which will be discussed later) why Einstein's proposal

$$\mathfrak{J}^\alpha = \varepsilon^{\alpha\beta\gamma\delta} g_{[[\beta\gamma],\delta]} \tag{3}$$

may not be valid. Of course, should we adopt

$$\mathfrak{J}^\alpha = \mathfrak{g}^{[\alpha,\beta]}\Gamma_\beta \tag{4}$$

from chapter 2, then electrodynamics could not be Maxwellian. Indeed, it would bear some resemblance to the theory considered by Dirac (in 1952) although he had current directly proportional to the potential. But we can be more specific.

Let us recall that there are two distinct formulations of Maxwell's equations. If $f_{\mu\nu}$ is defined by

$$f_{ij} = B_k, \quad f_{k0} = E_k, \tag{5}$$

where E is the electric intensity vector, B the magnetic induction, and (i, j, k) a cyclic permutation of $(1, 2, 3)$, then

$$f^{\mu\nu}{}_{;\nu} = J^\mu, \quad f_{[\mu\nu,\lambda]} = 0, \tag{6}$$

ignoring the fact that the material tensor $h^{\mu\nu}$ may be distinct (or even independent, as in Mie's theory) from $f^{\mu\nu}$. However, if

$$f_{ij} = E_k, \quad f_{0k} = B_k, \tag{7}$$

then

$$f^{\mu\nu}{}_{;\nu} = 0, \quad f_{[\mu\nu,\lambda]} = \varepsilon_{\mu\nu\lambda\sigma}\mathfrak{J}^\sigma. \tag{8}$$

In (6), J^μ is the current vector and, in (8), $\mathfrak{J}^\mu$ is the current vector density. I shall discuss the problem of geometrical symmetry in the next chapter, but some standard results can be anticipated. Thus, for a static, spherically symmetric field, the tensor $R_{[\mu\nu]}$, satisfying the equation

$$R_{[[\mu\nu],\,\lambda]} = 0,$$

has the general form

$$R_{[23]} = W\sin\theta, \quad R_{[10]} = p(r), \tag{9}$$

where W is a constant and $p(r)$ an arbitrary function of r. The corresponding components of $g_{\mu\nu}$ are given by

$$g_{(\mu\nu)} = \mathrm{diag}\,(\gamma, -\alpha, -\beta, -\beta\sin^2\theta), \quad g_{[23]} = f\sin\theta, \quad g_{[10]} = w, \tag{10}$$

where α, β, γ, f, and w are functions of r only. Vanishing of w is a sufficient but not a necessary condition for p to be zero, and the same is true for f and W. Now, Papapetrou's solution when

$$f = 0, \quad w \neq 0, \tag{11}$$

is

$$\alpha^{-1} = \left(1 - \frac{2m}{r}\right), \quad \alpha\gamma = \left(1 + \frac{l^2}{r^4}\right), \quad w = \pm\frac{l^2}{r^2},$$

$$R_{[10]} = -\frac{2}{r}\frac{\mathrm{d}}{\mathrm{d}r}\left(\frac{w}{\alpha}\right) - \frac{2}{r^2}\frac{w}{\alpha}, \tag{12}$$

where m and l^2 are real constants. Then both $R_{[\mu\nu]}$ and $w_{\mu\nu}$ describe fields of order r^{-4} and therefore can not represent a simple point charge. On the other hand, if f_{23} is identified as the electric field then

$$R_{[23]} = (W/r^2)(\hat{r}) \tag{13}$$

and we obtain the classical Coulomb field with W as charge in electrostatic units. Since $R_{[\mu\nu]}$ is the curl of a vector, the first form of Maxwell's equations, given by equations (6) rather than that of equations (8), is indicated. There are two ways in which a contradiction can be avoided. We can assume that the electrodynamics in empty space (with $J^\mu = 0$) only is given by GFT. Then the two forms of Maxwell's equations are interchangeable and both the material and intensity tensors are represented by curls. In this case however, we should have to abandon the Coulomb interpretation (13) as well as all hope of arriving at a cosmological interpretation of space–time. Moreover, we should be unable to avoid a contradiction later on. Hence, I prefer the second alternative, which is to conclude that the electrodynamics of GFT is like that of the Born–Infeld nonlinear theory. The induction $p^{\mu\nu}$ and the intensity $s_{\mu\nu}$ tensors then satisfy the equations

$$(\sqrt{(-h)}p^{\mu\nu})_{,\nu} = 0 = s_{[\mu\nu,\lambda]}, \tag{14}$$

where $h = \det(g_{(\mu\nu)})$, and the first problem resolves itself.

8. Equations of motion in a cylindrical field

The EIH method can be extended to the case of motion in a cylindrical field if certain assumptions are allowed, and I conclude this chapter with a brief indication of the results.

Consider the motion, say, of an electron in the field of a line charge distribution. The situation is complicated by the fact that the field of the former is spherically symmetric (or so we can expect), and that of the latter cylindrically symmetric. However, the line charge can be modelled approximately as a linear distribution of discrete point charges and then the calculations of section 3 apply again. Thus the electromagnetic force is given by

$$-\frac{1}{2\pi}\int^{s} S_{ij}n^{j}\mathrm{d}S, \tag{1}$$

with

$$S_{ij} = \tfrac{1}{4}\delta_{ij}\varphi^{2} - \tfrac{1}{2}\varphi_{,i}\Phi_{,j} + \tfrac{1}{2}\varphi_{,ij}\Phi, \tag{2}$$

where φ is the classical potential and

$$\Phi_{,ss} = \varphi. \tag{3}$$

Since equation (3) is linear, we can assume that at least up to λ^{4} the solutions are additive, and consider the spherical and cylindrical cases separately.

Let ρ_{1} be the radial distance of a field point from the moving charge, and ρ_{2} its distance from the line charge. We solve separately the spherical equation

$$\frac{\mathrm{d}^{2}\Phi_{1}}{\mathrm{d}\rho_{1}^{2}} + \frac{2}{\rho_{1}}\frac{\mathrm{d}\Phi_{1}}{\mathrm{d}\rho_{1}} = \frac{k_{1}}{\rho_{1}}, \tag{4}$$

and the cylindrical

$$\frac{\mathrm{d}^{2}\Phi_{2}}{\mathrm{d}\rho_{2}^{2}} + \frac{1}{\rho_{2}}\frac{\mathrm{d}\Phi_{2}}{\mathrm{d}\rho_{2}} = k_{2}\ln\rho_{2}, \tag{5}$$

where k_{1} and k_{2} are constants. The complete solution is

$$\Phi = \Phi_{1} + \Phi_{2} = -\frac{a_{1}}{\rho_{1}} + \frac{k_{1}}{2}\rho_{1} + b_{1}$$
$$+ \frac{k_{2}}{4}[(\rho_{2}^{2} + a)\ln\rho_{2} - (\rho_{2}^{2} + b)], \tag{6}$$

where a, b, a_1 and b_1 are constants of integration. We must add to Φ a function $\Phi^*(\rho_1, \rho_2)$ which we suppose to be separately harmonic in ρ_1 and ρ_2. Then

$$\Phi^* = -\frac{L}{\rho_1} \ln \frac{\rho_2}{\rho_0} + M \ln \frac{\rho_2}{\rho_0} \tag{7}$$

in which L and ρ_0 again are constants and we can take $M = 0$ since this term is already contained in Φ.

The only surviving term in the surface integral (1) is

$$\frac{k_1 k_2}{6\pi} \left[2\rho_2 \ln \rho_2 + \left(1 + \frac{a}{\rho_2^2} \right) \rho_2 - 2\rho_2 \right] - \frac{2k_2}{3\rho_2} \left[a_1 + L \ln \frac{\rho_2}{\rho_0} \right], \tag{8}$$

in the radial direction away from the line distribution. For a test particle we take $k_1 = 0$ so that the force becomes

$$-\frac{2k_2}{3\rho_2} \left[a_1 + L \ln \frac{\rho_2}{\rho_0} \right]. \tag{9}$$

We may observe that because the force is radial (which is not surprising for a quasi-stationary approximation), the cross term $f \wedge B$ in the Lorentz force is definitely missing. For this reason, the case considered here is not likely to be realistic or for that matter suitable as an empirical test of the theory.

5 Solution of the field equations

1. The Tonnelat affine connection

I have outlined in the preceding chapters the overall structure of generalised field theory and the solution of the problem of motion which, specifically, gave an indication of how the theory should be interpreted. However, the Einstein–Infeld–Hoffman method used in obtaining the equations of motion of a test particle is not entirely free from doubt (as to its physical meaning; I shall comment on this point in a later chapter) and can not alone be responsible for the physical interpretation of the theory. Indeed, the latter can only be derived from detailed knowledge of the exact solutions of the field equations. Unfortunately, only very few of these have been found and only in the case of static, spherical symmetry is the general solution known. Not surprisingly therefore, it is this solution which forms the basis of most of the predictions which can be made at present. It was found by Vanstone (1962).

Before discussing Vanstone's solution, let us consider the purely algebraic problem of determining the affine connection in terms of the field tensor $g_{\mu\nu}$ and of its first derivatives, corresponding to the determination of the Christoffel brackets of general relativity. This has been discussed by Hlavaty (1957; it is, of course, one of my main claims that Hlavaty's book is wrongly entitled: the geometry of the non-symmetric theory is just as Riemannian as that of GR) but the method followed herein is due to Mme Tonnelat and has been modified only slightly by Gregory and myself. The aim is to solve for $\tilde{\Gamma}^{\lambda}{}_{\mu\nu}$ the sixty-four equations

$$g_{\mu\nu,\lambda} - (\Gamma^{\sigma}{}_{\mu\lambda} + \tfrac{2}{3}\delta^{\sigma}{}_{\mu}\Gamma_{\lambda})g_{\sigma\nu} - (\Gamma^{\sigma}{}_{\lambda\nu} + \tfrac{2}{3}\delta^{\sigma}{}_{\lambda}\Gamma_{\nu})g_{\mu\sigma} = 0 . \tag{1}$$

It is convenient to revert to the notation

$$g_{(\mu\nu)} = h_{\mu\nu}, \quad g_{[\mu\nu]} = k_{\mu\nu}, \tag{2}$$

with the determinants

$$h < 0, \quad k > 0. \tag{3}$$

Then

$$\Gamma^{\lambda}{}_{\mu\nu} = \left\{ {\lambda \atop \mu\nu} \right\}_h + u^{\lambda}{}_{\mu\nu}, \tag{4}$$

where

$$h_{\rho\lambda}u^{\lambda}{}_{\mu\nu} = \Gamma^{\sigma}{}_{[\mu\rho]}k_{\sigma\nu} + \Gamma^{\sigma}{}_{[\nu\rho]}k_{\sigma\mu} - \tfrac{1}{3}\Gamma_{\mu}g_{\rho\nu} - \tfrac{1}{3}\Gamma_{\nu}g_{\rho\mu}$$
$$= u_{\mu\nu:\rho}, \tag{5}$$

say. It can be shown easily also that

$$\Gamma_{[\mu\nu]:\rho} \equiv h_{\rho\lambda}\Gamma^{\lambda}{}_{[\mu\nu]} = -\tfrac{1}{2}k_{\mu\nu\rho} + \nabla_{\rho}k_{\mu\nu} - (u^{\sigma}{}_{\mu\rho}k_{\sigma\nu} - u^{\sigma}{}_{\nu\rho}k_{\sigma\mu})$$
$$+ \tfrac{1}{3}\Gamma_{\mu}g_{\nu\rho} - \tfrac{1}{3}\Gamma_{\nu}g_{\mu\rho} - \tfrac{2}{3}\Gamma_{\rho}k_{\mu\nu}, \tag{6}$$

where

$$k_{\mu\nu\rho} = k_{[\mu\nu,\rho]} \tag{7}$$

and ∇_{ρ} is the covariant differentiation operator with the h-Christoffel brackets as affine connection. We should notice that in the expressions $u_{\mu\nu:\rho}$ and $\Gamma_{[\mu\nu]:\rho}$ the colon does not indicate a differentiation. Similarly to the notation of (7), we shall also use the quantities

$$u_{\mu\nu\rho} = u_{[\mu\nu:\rho]}, \quad \Gamma_{\mu\nu\rho} = \Gamma_{[[\mu\nu]:\rho]}. \tag{8}$$

Then

$$u_{\mu\nu\rho} = -\tfrac{2}{3}\Gamma_{[\mu}h_{\nu\rho]}, \quad \Gamma_{\mu\nu\rho} = -\tfrac{1}{2}k_{\mu\nu\rho},$$

as may be readily verified.

The algebraic problem now reduces to sorting out equation (6), or rather, of introducing purely tensorial operations which enable us to construct from it enough independent relations with scalar coefficients for a solution to be written down. One of these operations is the well-known duality relation

$$A^{*}{}_{\dots\mu\nu\dots} = \tfrac{1}{2}\sqrt{(-h)}\varepsilon_{\mu\nu\alpha\beta}h^{\alpha\rho}h^{\beta\sigma}A_{\dots\rho\sigma\dots}, \tag{9}$$

on a skew pair of indices. The second we call, following Tonnelat, the 'barring' of an index. It consists of raising a covariant index with $h^{\mu\nu}$ and subsequently lowering it with $k_{\mu\nu}$ and can be repeatedly applied so that

$$A^{\dots}{}_{\dots\bar{\lambda}\dots} = k_{\lambda\alpha}h^{\alpha\beta}A^{\dots}{}_{\dots\beta\dots}, \tag{10}$$

$$A^{\dots}{}_{\dots\bar{\bar{\lambda}}\dots} = k_{\lambda\alpha}h^{\alpha\beta}k_{\beta\rho}h^{\rho\sigma}A^{\dots}{}_{\dots\sigma\dots}, \tag{11}$$

and so on.

Since

$$\sqrt{k} = \tfrac{1}{8}\varepsilon^{\mu\nu\rho\sigma}k_{\mu\nu}k_{\rho\sigma}, \tag{12}$$

and

$$g = h + k + \tfrac{1}{2}h^{\mu\rho}h^{\nu\sigma}k_{\mu\nu}k_{\rho\sigma}, \tag{13}$$

from the general theory of determinants, it can be shown readily that

$$A_{\bar{\bar{\rho}}} = w^2 A_{\rho} + (1 - j^2)A_{\bar{\rho}} \tag{14}$$

where

$$w = \sqrt{(-k/h)}, \quad j^2 = (g-k)/h. \tag{15}$$

There exists also a useful relation

$$2h^{\sigma\lambda}k_{\lambda\rho}k^*{}_{\sigma[\mu}A_{\nu]} = -\sqrt{(-h)}h^{\sigma\lambda}\varepsilon_{\mu\nu\bar{\rho}\sigma}A_\lambda, \tag{16}$$

duality star referring to the index pair (μ, ν). Let us write

$$\bar{\Gamma}_{\mu\nu\rho} = \Gamma_{[[\mu\nu]:\bar{\rho}]}, \quad \bar{\bar{\Gamma}}_{\mu\nu\rho} = \Gamma_{[[\mu\nu]:\bar{\rho}]}, \tag{17}$$

the bars remaining in position under a cyclic summation. Then we find that

$$\Gamma_{[\mu\nu]:\rho} - h^{\sigma\lambda}(\bar{\Gamma}_{\mu\lambda\rho}k_{\sigma\nu} - \bar{\Gamma}_{\nu\lambda\rho})k_{\sigma\mu})$$
$$\quad - 2h^{\sigma\lambda}(\Gamma_{[\rho\lambda]:\bar{\mu}}k_{\sigma\nu} - \Gamma_{[\rho\lambda]:\bar{\nu}}k_{\sigma\mu}) + \Gamma_{[\mu\nu]:\bar{\bar{\rho}}} - \bar{\bar{\Gamma}}_{\mu\nu\rho}$$
$$= -\tfrac{1}{2}k_{\mu\nu\rho} + \nabla_\rho k_{\mu\nu} + \tfrac{1}{3}[\Gamma_\mu(h_{\nu\rho} + h^{\sigma\lambda}k_{\lambda\rho}\,k_{\sigma\nu})$$
$$\qquad\qquad\qquad - \Gamma_\nu(h_{\mu\rho} + h^{\sigma\lambda}k_{\lambda\rho}k_{\sigma\mu})]. \tag{18}$$

We now note that

$$\Gamma^\sigma{}_{(\rho\sigma)} = \partial_\rho \ln\sqrt{(-g)} - \tfrac{5}{3}\Gamma_\rho,$$

and that, whether

$$k = 0$$

or not, we can define

$$k^{\mu\nu} = \frac{1}{2\sqrt{k}}\varepsilon^{\mu\nu\alpha\beta}k_{\alpha\beta}, \quad k_{\mu\nu} = \tfrac{1}{2}\sqrt{k}\,\varepsilon_{\mu\nu\alpha\beta}k^{\alpha\beta}.$$

A somewhat lengthy but essentially straightforward calculation the details of which are given in Mme Tonnelat's article, yields from equations (6) and (18)

$$\Gamma_{[\mu\nu]:\rho} - 2w\Gamma^*{}_{[\mu\nu]:\rho} + \Gamma_{[\mu\nu]:\bar{\bar{\rho}}} = J_{[\mu\nu]:\rho} + W_{[\mu\nu]:\rho}, \tag{19}$$

where

$$J_{[\mu\nu]:\rho} = -\tfrac{1}{2}k_{\mu\nu\rho} + \nabla_\rho k_{\mu\nu} + \tfrac{1}{2}wk^*{}_{\mu\nu\rho} + \tfrac{1}{4}wk^*{}_{\mu\nu}k^{\sigma\tau}k_{\sigma\tau\rho}$$
$$\quad - (k_{\mu\nu}\delta^\lambda{}_\rho - \tfrac{1}{2}\sqrt{k}\,\varepsilon_{\mu\nu\rho\sigma}k^{\lambda\sigma})\partial_\lambda \ln(j^2 - w^2)$$
$$\quad - \tfrac{1}{2}w(\varepsilon^*{}_{\mu\nu\rho\sigma}k^{\lambda\sigma} - k^*{}_{\mu\nu}\delta^\lambda{}_\rho)\partial_\lambda \ln\left(\frac{j^2 - w^2}{w^2}\right), \tag{20}$$

and

$$W_{[\mu\nu]:\rho} = \tfrac{2}{3}h^{\sigma\lambda}[k\varepsilon_{\mu\nu\rho\sigma}\Gamma_\lambda + k_{\lambda\rho}k_{\sigma[\mu}\Gamma_{\nu]}] + \tfrac{2}{3}\Gamma_{[\mu}h_{\nu]\rho}. \tag{21}$$

If we now apply to the equation (19) the star (duality) operation, the double-bar operation, and the two operations combined, in succession, and recall the identity (14) as well as

$$A^{**}{}_{\ldots\mu\nu\ldots} = -A_{\ldots\mu\nu\ldots},$$

which is self-evident, we obtain in addition to it the equations

$$2w\Gamma_{[\mu\nu]:\rho} + \Gamma^*{}_{[\mu\nu]:\rho} + \Gamma^*{}_{[\mu\nu]:\bar{\bar{\rho}}} = J^*{}_{[\mu\nu]:\rho} + W^*{}_{[\mu\nu]:\rho} \tag{22}$$

$$w^2\Gamma_{[\mu\nu]:\rho} + (2 - j^2)\Gamma_{[\mu\nu]:\bar{\rho}} - 2w\Gamma^*_{[\mu\nu]:\bar{\rho}} = J_{[\mu\nu]:\bar{\rho}} + W_{[\mu\nu];\bar{\rho}}, \quad (23)$$

and

$$w^2\Gamma^*_{[\mu\nu]:\rho} + 2w\Gamma_{[\mu\nu]:\bar{\rho}} + (2 - j^2)\Gamma^*_{[\mu\nu]:\bar{\rho}} = J^*_{[\mu\nu]:\bar{\rho}} + W^*_{[\mu\nu]:\bar{\rho}}. \quad (24)$$

If we also write

$$A_{\mu\nu:\lambda} = -\tfrac{2}{3}\Gamma_\nu h_{\mu\lambda}, \quad (25)$$

then it can be verified that

$$A_{[\mu\nu]:\lambda} - 2wA^*_{[\mu\nu]:\lambda} + A_{[\mu\nu]:\bar{\lambda}} = W_{[\mu\nu]:\lambda}. \quad (26)$$

Hence, the solution of the four equations (19), (22), (23) and (24) can be written as

$$(a^2 + b^2)\Gamma_{[\mu\nu]:\rho} = U_{[\mu\nu]:\rho} + \tfrac{2}{3}(a^2 + b^2)\Gamma_{[\mu} h_{\nu]\rho}, \quad (27)$$

where

$$U_{[\mu\nu]:\rho} = (2wb - ac)J_{[\mu\nu]:\rho} + (2wa + b)J^*_{[\mu\nu]:\rho}$$
$$+ aJ_{[\mu\nu]:\bar{\rho}} - bJ^*_{[\mu\nu]:\bar{\rho}}, \quad (28)$$

and

$$a = 5w^2 + j^2 - 2, \quad b = 2w(3 - j^2), \quad c = 2 - j^2. \quad (29)$$

Thus, provided that

$$a \neq 0 \neq b, \quad (30)$$

$\Gamma_{[\mu\nu]:\rho}$ is determined by (27) in terms of $g_{\mu\nu}$, its first derivatives and Γ_μ. Then $\Gamma^\lambda_{\mu\nu}$ is similarly determined from equations (4) and (5). But it can be shown easily, that if

$$g \neq 0, \quad (31)$$
$$U_\mu = h^{\nu\rho} U_{[\mu\nu]:\rho} = 0 \quad (32)$$

so that Γ_μ can not be determined from equation (27) and therefore also not from the equation (1) whose rank is now seen to be 60 instead of 64. This justifies explicitly my use of Γ_μ as a 'physical' quantity to be determined by the differential field equations instead of by the arbitrary (within reason) algebraic ones.

The conditions (30) and (31) for the algebraic solution to be unique can be combined into Hlavaty's

$$(g/h)(g/h - 2) \neq 0. \quad (33)$$

2. On geometrical symmetry

The Tonnelat solution for the affine connection ($\tilde{\Gamma}^\lambda_{\mu\nu}$ of Schrödinger, which alone can be uniquely determined by algebraic processes) is, because of its complexity, of little use in finding general solutions of the field equations. I have recorded it as a check on whatever we may be able to obtain otherwise. And, as in general relativity, exact solutions can be found only under restricted symmetry requirements. The situation in the nonsymmetric theory however is quite different from in GR. There, definition of geometrical symmetry did not lead to serious problems for two reasons. First, a Riemannian space–time can always be embedded in a higher-dimensional (pseudo-) Euclidean space in which intuitive concepts of symmetry apply and the results can be projected back into the Riemannian manifold. Secondly, the latter is identified by the strong principle of geometrisation with the effects of a gravitational field. Hence, any symmetry restriction on the space–time becomes automatically an equivalent condition on the field.

In GFT on the other hand, the field equations are written in terms of the field ($g_{\mu\nu}$ and Γ_μ) and the affine connection and only after they have been solved can the (Riemannian) geometry of the space–time be determined from the metric hypothesis equations. It follows that, at any rate to start with, symmetry restrictions under which the field equations are solved apply to the physical fields rather than to the background geometry. An attempt to define a more general concept of symmetry with immediate geometrical implications will be described in the next section.

At the moment, we can lay down only a formal definition of symmetry from the fact that the group $\mathcal{G}$ of coordinate transformations

$$x'^\lambda = x'^\lambda(x^\mu) \tag{1}$$

with a nonvanishing Jacobian (a.e.) is the same in GFT as in GR (the principle of covariance).

Let $y(x^\lambda)$ be a mathematical object, possibly the space–time manifold itself, on which an element of $\mathcal{G}$ induces the transformation

$$y'(x') = y(x'). \tag{2}$$

Then this element is said to be a symmetry of y and a subgroup of $\mathcal{G}$ which is the set of all such elements is called the symmetry group of y. This, of course, is very vague since we have not specified the nature of y beyond the requirement that elements of $\mathcal{G}$ (coordinate transformations) should induce some transformation law on it.

Suppose now that a tensor $A_{\mu\nu}$ possesses a symmetry under an infinitesimal transformation

$$x'^\lambda = x^\lambda + \varepsilon\xi^\lambda, \quad \varepsilon^2 \ll \varepsilon, \tag{3}$$

where ξ^λ is a known function of the coordinates and which belongs to $\mathcal{G}$. The transformation (3) is a mapping of the manifold onto itself.

From the tensor law of transformation

$$A'_{\mu\nu}(x') = A_{\mu\nu}(x) - \varepsilon\xi^\sigma{}_{,\mu}A_{\sigma\nu} - \varepsilon\xi^\sigma{}_{,\nu}A_{\mu\sigma} + O(\varepsilon^2), \tag{4}$$

and by Taylor expansion

$$A'_{\mu\nu}(x') = A'_{\mu\nu}(x) + \varepsilon A'_{\mu\nu,\sigma}\xi^\sigma + O(\varepsilon^2). \tag{5}$$

Hence, from the formal definition of symmetry above,

$$A_{\mu\nu,\sigma}\xi^\sigma + A_{\sigma\nu}\xi^\sigma{}_{,\mu} + A_{\mu\sigma}\xi^\sigma{}_{,\nu} = 0. \tag{6}$$

When $A_{\mu\nu}$ is replaced by the metric tensor of a Riemannian space–time, equation (6) gives the standard Killing equations. If on the other hand, we apply the transformation to the equation

$$g_{\mu\nu,\lambda} - \tilde{\Gamma}^\sigma{}_{\mu\lambda}g_{\sigma\nu} - \tilde{\Gamma}^\sigma{}_{\lambda\nu}g_{\mu\sigma} = 0,$$

we find that

$$g_{\mu\rho}(\xi^\rho{}_{,\nu} + \tilde{\Gamma}^\rho{}_{\sigma\nu}\xi^\sigma) + g_{\sigma\nu}(\xi^\sigma{}_{,\mu} + \tilde{\Gamma}^\rho{}_{\mu\sigma}\xi^\sigma) = 0$$

or

$$g_{\mu\sigma}\xi^{\sigma+}{}_{;\nu} + g_{\sigma\nu}\xi^{\sigma-}{}_{;\mu} = 0. \tag{7}$$

This result is useful when we are discussing the symmetry of a geometrical manifold but not when our concern is the purely physical field. To obtain forms of a $g_{\mu\nu}$ which are sufficiently simplified for us to be able to solve the corresponding field equations, we must look to more primitive ideas of symmetry under transformations which we may recognise as being equivalent to them.

Let us consider first the case of a field with general axial symmetry and the cartesian z axis as the axis of symmetry. Let $c_{\mu\nu}$ denote the fundamental tensor expressed in cartesian coordinates. We expect it to be unchanged under the condition of axial symmetry by the reflection

$$x' = -x, \quad y' = -y, \quad z' = z, \quad t' = t \tag{8}$$

whence, from the transformation law of $c_{\mu\nu}$, we readily find that

$$c_{\mu\nu} = \begin{bmatrix} c_1 & 0 \\ 0 & c_2 \end{bmatrix} \tag{9}$$

where 2×2 matrices c_1 and c_2 are quite general:

$$c_1 = \begin{bmatrix} c_{11} & c_{12} \\ c_{21} & c_{22} \end{bmatrix}, \quad c_2 = \begin{bmatrix} c_{33} & c_{34} \\ c_{43} & c_{44} \end{bmatrix}. \tag{10}$$

Let now

$$x = r\sin\theta\cos\varphi, \quad y = r\sin\theta\sin\varphi, \quad z = r\cos\theta, \quad t = \tau. \tag{11}$$

The transformation matrix

$$\left[\frac{\partial x^\mu}{\partial x'^\nu}\right] = \begin{bmatrix} \alpha & \beta \\ \gamma & \delta \end{bmatrix}, \tag{12}$$

where

$$\alpha = \begin{bmatrix} \sin\theta\cos\varphi & \sin\theta\sin\varphi \\ r\cos\theta\cos\varphi & r\cos\theta\sin\varphi \end{bmatrix}, \quad \beta = \begin{bmatrix} \cos\theta & 0 \\ -r\sin\theta & 0 \end{bmatrix},$$

$$\gamma = \begin{bmatrix} -r\sin\theta\sin\varphi & r\sin\theta\cos\varphi \\ 0 & 0 \end{bmatrix},$$

and

$$\delta = \begin{bmatrix} 0 & 0 \\ 0 & 1 \end{bmatrix}.$$

Then

$$g_{(\mu\nu)} = \begin{bmatrix} \alpha & \beta \\ \gamma & \delta \end{bmatrix} \begin{bmatrix} c_1 & 0 \\ 0 & c_2 \end{bmatrix} \begin{bmatrix} \alpha^{\mathrm{T}} & \gamma^{\mathrm{T}} \\ \beta^{\mathrm{T}} & \delta^{\mathrm{T}} \end{bmatrix}$$
$$= \begin{bmatrix} \alpha c_1\alpha^{\mathrm{T}} + \beta c_2\beta^{\mathrm{T}} & \alpha c_1\gamma^{\mathrm{T}} + \beta c_2\delta^{\mathrm{T}} \\ \gamma c_1\alpha^{\mathrm{T}} + \delta c_2\beta^{\mathrm{T}} & \gamma c_1\gamma^{\mathrm{T}} + \delta c_2\delta^{\mathrm{T}} \end{bmatrix}, \tag{13}$$

where

$$\alpha c_1\alpha^{\mathrm{T}} = Q_1 \begin{bmatrix} \sin^2\theta & r\sin\theta\cos\theta \\ r\sin\theta\cos\theta & r^2\cos^2\theta \end{bmatrix},$$

$$\beta c_2\beta^{\mathrm{T}} = c_{33} \begin{bmatrix} \cos^2\theta & -r\sin\theta\cos\theta \\ -r\sin\theta\cos\theta & r^2\sin^2\theta \end{bmatrix},$$

$$\alpha c_1\gamma^{\mathrm{T}} = Q_2 \begin{bmatrix} \sin\theta & 0 \\ r\cos\theta & 0 \end{bmatrix} r\sin\theta,$$

$$\beta c_2\delta^{\mathrm{T}} = c_{34} \begin{bmatrix} 0 & \cos\theta \\ 0 & -r\sin\theta \end{bmatrix},$$

$$\gamma c_1\alpha^{\mathrm{T}} = Q_3 \begin{bmatrix} \sin\theta & r\cos\theta \\ 0 & 0 \end{bmatrix} r\sin\theta,$$

$$\delta c_2\beta^{\mathrm{T}} = c_{43} \begin{bmatrix} 0 & 0 \\ \cos\theta & -r\sin\theta \end{bmatrix},$$

$$\gamma c_1\gamma^{\mathrm{T}} = Q_4 \begin{bmatrix} 1 & 0 \\ 0 & 0 \end{bmatrix} r^2\sin^2\theta,$$

$$\delta c_2\delta^{\mathrm{T}} = \begin{bmatrix} 0 & 0 \\ 0 & c_{44} \end{bmatrix}$$

$$\tag{14}$$

and

$$Q_1 = c_{11}\cos^2\varphi + (c_{12} + c_{21})\sin\varphi\cos\varphi + c_{22}\sin^2\varphi,$$
$$Q_2 = (c_{22} - c_{11})\sin\varphi\cos\varphi + c_{12} - (c_{12} + c_{21})\sin^2\varphi,$$
$$Q_3 = (c_{22} - c_{11})\sin\varphi\cos\varphi + c_{21} - (c_{12} + c_{21})\sin^2\varphi,$$
$$Q_4 = c_{11}\sin^2\varphi - (c_{12} + c_{21})\sin\varphi\cos\varphi + c_{22}\cos^2\varphi.$$
(15)

Now the (pseudo-) Killing equations (7) demand that the fundamental tensor should be φ independent for general axial symmetry, whence equations (15) give

$$c_{12} = -c_{21}, \quad c_{11} = c_{22} = c.$$
(16)

Then

$$g_{\mu\nu} = \begin{bmatrix} c\sin^2\theta + c_{33}\cos^2\theta & (c - c_{33})r\sin\theta\cos\theta & c_{12}r\sin^2\theta & c_{34}\cos\theta \\ (c - c_{33})r\sin\theta\cos\theta & r^2(c\cos^2\theta + c_{33}\sin^2\theta) & c_{12}r^2\sin\theta\cos\theta & -c_{34}r\sin\theta \\ -c_{12}r\sin^2\theta & -c_{12}r^2\sin\theta\cos\theta & cr^2\sin^2\theta & 0 \\ c_{43}\cos\theta & -c_{43}r\sin\theta & 0 & c_{44} \end{bmatrix}$$
(17)

Hence, the general axially symmetric, skew fundamental tensor has the form

$$g_{[\mu\nu]} = \begin{bmatrix} 0 & 0 & u\sin^2\theta & w\cos\theta \\ 0 & 0 & ur\sin\theta\cos\theta & -wr\sin\theta \\ -u\sin^2\theta & -ur\sin\theta\cos\theta & 0 & 0 \\ -w\cos\theta & wr\sin\theta & 0 & 0 \end{bmatrix}$$
(18)

where $u = rc_{12}$ and $w = c_{[34]}$ are functions of r, θ and t only. The general, spherically symmetric form of the fundamental tensor can be obtained from (17) by applying to it the permissible transformation

$$\begin{bmatrix} r & 0_3 \\ 0_3 & I_3 \end{bmatrix}$$

followed by a rotation

$$\begin{bmatrix} \cos\theta & -\sin\theta & \\ \sin\theta & \cos\theta & 0_2 \\ 0_2 & & I_2 \end{bmatrix}$$

The result is

$$g_{\mu\nu} = \begin{bmatrix} -\alpha & 0 & 0 & h+w \\ 0 & -\beta & u\sin\theta & 0 \\ 0 & -u\sin\theta & -\beta\sin^2\theta & 0 \\ h-w & 0 & 0 & \gamma \end{bmatrix},$$
(19)

where

$$\alpha = -r^2 c_{33}, \quad \beta = r^2 c, \quad \gamma = c_{44}, \quad u = r^2 c_{12},$$
$$h = rc_{(34)}, \quad w = rc_{[34]}$$

$$(20)$$

are functions of r and t only.

Similarly, for a cylindrical coordinate system

$$(r, z, \theta, t),$$

(21)

instead of (11), we obtain a cylindrically symmetric fundamental tensor

$$g_{\mu\nu} = \begin{bmatrix} -\alpha & 0 & 0 & 0 \\ 0 & -\alpha & e & 0 \\ 0 & -e & -\beta & -h \\ 0 & 0 & -h & \gamma \end{bmatrix},$$

(22)

with α, β, γ, e and h functions of the radial distance r from the axis of symmetry (and of t for a nonstatic case). These are the only two cases which will be discussed in the present monograph (and only the first is solved).

In the above, and contrary to the convention adopted elsewhere in this monograph, I have written $x^4 = t$ rather than $x^0 = t$ to make reference to the work of Papapetrou, who first found the form (19) in 1948, easier.

3. On universal symmetry of Gregory

A geometric object y (x), where x is a point of the space–time whose coordinates (in some coordinate system) are x^λ, is said to be form-invariant under a transformation (2.1) if $y'(x')$ is the same function of the transformed arguments as $y(x)$ was of x^{λ}'s. The condition that y should be form-invariant under an infinitesimal transformation (2.3) is that the Lie derivative of y with respect to ξ^λ should vanish:

$$\mathscr{L}_\xi(y) = 0.$$

(1)

The requirement of form-invariance in a Riemannian manifold coincides with the definition of symmetry introduced in the preceding section. In particular, if the metric tensor is symmetric with respect to some subgroup of the general Lie group $\mathscr{G}$ of allowable transformations of coordinates, all geometrical quantities defined in the manifold, that is the Christoffel brackets, Riemann and Ricci tensors and so on, will be likewise symmetric. Hence also any solutions of the general relativistic field equations will exhibit the same symmetries as may be imposed *a priori* on the geometry in order to derive them in the first instance. It should be pointed out, however, that in the case of a theory such as Einstein–Maxwell, the symmetry of the electromagnetic field must be assumed independently of that of the metric tensor (gravitational field).

Similarly, in GFT, symmetry of the field $(g_{\mu v})$ does not guarantee that of the space–time $(a_{\mu v}$ defined by the metric hypothesis). Indeed, if $g_{\mu v}$ is symmetric, that is form-invariant under a symmetry subgroup of $\mathscr{G}$, in some sense, then

$$\mathscr{L}_\xi(\tilde{\Gamma}^\lambda{}_{\mu v}) = 0, \tag{2}$$

whence

$$\mathscr{L}_\xi(\tilde{\Gamma}^\lambda{}_{(\mu v)}) = \mathscr{L}_\xi\left(\left\{ \begin{matrix} \lambda \\ \mu v \end{matrix} \right\}_a\right) = 0. \tag{3}$$

Therefore the Riemannian background (which is the physical space–time) allows an affine motion ξ^λ but this need not be a motion of the space since we may have

$$\mathscr{L}_\xi(a_{\mu v}) \neq 0. \tag{4}$$

However, we have not taken into account so far the full set of the field equations. It may turn out in practice that the latter provide conditions for the Lie derivative of the metric tensor $a_{\mu v}$ to vanish for particular ζ^λ (that is for a particular symmetry group). In that case all the geometric (and physical!) objects defined in the space–time are form-invariant with respect to the infinitesimal transformations generated by that ξ.

We then say that ξ^λ is a universal symmetry of the space–time. For example, the latter will be universally spherically symmetric if it admits in polar coordinates (t, r, θ, φ) the vectors

$$\overset{1}{\xi}{}^\lambda = (0, 0, \sin\varphi, \cot\theta\cos\varphi),$$

$$\overset{2}{\xi}{}^\lambda = (0, 0, -\cos\varphi, \cot\theta\sin\varphi),$$

$$\overset{3}{\xi}{}^\lambda = (0, 0, 0, -1), \tag{5}$$

as universal symmetries. The space will also be static if in addition it admits the vector

$$\overset{4}{\xi}{}^\lambda = (1, 0, 0, 0). \tag{6}$$

The (universal) symmetry will be preserved by the group of transformations given by

$$\left. \begin{matrix} r' = at + f(r), \\ t' = bt + g(r), \end{matrix} \right\} \tag{7}$$

where a, b are constants and f and g are arbitrary functions of r subject only to the condition

$$b\frac{\mathrm{d}f}{\mathrm{d}r} - a\frac{\mathrm{d}g}{\mathrm{d}r} \neq 0. \tag{8}$$

In this case the coordinates can always be found so that the symmetric part of $g_{\mu\nu}$ in (2.19) is diagonal:

$$h = 0. \tag{9}$$

It is this form of $g_{\mu\nu}$ that we shall now use to find the general, static spherically symmetric solution of the weak field equations.

4. The Vanstone solution

When the fundamental tensor is static and spherically symmetric and the coordinate system is given by

$$x^\lambda = (t, r, \theta, \varphi) \tag{1}$$

its nonvanishing components (with $g_{01} = 0$) are

$$\left. \begin{array}{l} g_{00} = \gamma, \quad g_{10} = w = -g_{01}, \quad g_{23} = u \sin\theta = -g_{32}, \\ g_{11} = -\alpha, \quad g_{22} = -\beta = g_{33}\operatorname{cosec}^2\theta, \end{array} \right\} \tag{2}$$

with α, β, γ, u and w functions of r only. Also

$$\alpha, \beta, \gamma$$

are strictly positive in order to preserve the general relativistic signature when

$$u = w = 0,$$

and $g_{\mu\nu}$ becomes the Schwarzschild metric of GR.

We also have

$$0 < -g = \alpha\gamma\left(1 - \frac{w^2}{\alpha\gamma}\right)(u^2 + \beta^2)\sin^2\theta, \tag{3}$$

so that

$$U = (1 - w^2/\alpha\gamma) > 0. \tag{4}$$

Let us also write

$$\rho^2 = u^2 + \beta^2 = \exp(2P), \quad y = \gamma u, x = \rho^2\alpha^{-1}, \quad \tan Q = \beta/u, \tag{5}$$

when, with dashes denoting derivatives with respect to r,

$$P' = \rho'/\rho \quad \text{and} \quad Q' = (\beta'u - \beta u')/(\beta^2 + u^2). \tag{6}$$

Then, either from the Tonnelat solution of section 1, or directly from the equations

$$g_{\mu\nu,\lambda} - \Gamma^\sigma{}_{\mu\lambda}g_{\sigma\nu} - \Gamma^\sigma{}_{\lambda\nu}g_{\mu\sigma} = 0.$$

(in which we have omitted the tilde from the affine connection, since for the time being we are not going to be interested in the vector potential) we obtain the nonvanishing components of the (Schrödinger) affine connection as follows.

$$
\left.
\begin{aligned}
&\Gamma^0{}_{(10)} = \tfrac{1}{2}y^{-1}y', \quad \Gamma^1{}_{00} = \tfrac{1}{2}\gamma\alpha^{-1}(\ln yU)', \\
&\Gamma^1{}_{[10]} = -2\Gamma^2{}_{[20]} = -2\Gamma^3{}_{[30]} \\
&\qquad = w\rho'/\alpha\rho = \tfrac{1}{2}\gamma w^{-1}(\ln U)', \\
&\Gamma^1{}_{11} = \tfrac{1}{2}\alpha^{-1}\alpha', \\
&\Gamma^1{}_{22} = \Gamma^1{}_{33}\operatorname{cosec}^2\theta = \tfrac{1}{2}\alpha^{-1}(uQ' - \beta P'), \\
&\Gamma^1{}_{[23]} = \tfrac{1}{2}\alpha^{-1}(\beta Q' + uP')\sin\theta_1 \\
&\Gamma^2{}_{33} = -\sin\theta\cos\theta, \quad \Gamma^2{}_{(12)} = \Gamma^3{}_{(13)} = \tfrac{1}{2}P', \\
&\Gamma^2{}_{(30)} = -\Gamma^3{}_{(20)}\sin^2\theta = \tfrac{1}{2}w\alpha^{-1}Q'\sin\theta, \\
&\Gamma^3{}_{(23)} = \cot\theta, \quad \text{and} \quad \Gamma^2{}_{[31]} = \Gamma^3{}_{[12]}\sin^2\theta = -\tfrac{1}{2}Q'\sin\theta.
\end{aligned}
\right\} \tag{7}
$$

It follows immediately that Γ_μ vanishes identically as required but

$$
\begin{aligned}
\mathfrak{g}^{[\mu\nu]} &= \sqrt{(-g)}g^{[\mu\nu]} \\
&= \rho\sqrt{(\alpha\gamma U)}\sin\theta
\begin{bmatrix}
0 & w/\alpha\gamma u & 0 & 0 \\
-w/\alpha\gamma u & 0 & 0 & 0 \\
0 & 0 & 0 & u/\rho^2\sin\theta \\
0 & 0 & u/\rho^2\sin\theta & 0
\end{bmatrix}
\end{aligned}
\tag{8}
$$

so that the equivalent equation

$$
\mathfrak{g}^{[\mu\nu]}{}_{,\nu} = 0
$$

gives

$$
\mathfrak{g}^{[01]}{}_{,1} = 0
$$

or

$$
w\rho = k\sqrt{(\alpha\gamma - w^2)}, \tag{9}
$$

where k is a real constant. We also have

$$
\frac{w^2}{\alpha\gamma} = 1 - U = \frac{k^2}{k^2 + \rho^2}, \quad U = \frac{\rho^2}{k^2 + \rho^2} \tag{10}
$$

With the components of the affine connection given by the equations (7), the only components of the Ricci tensor which do not vanish identically are

$$
\begin{aligned}
R_{11} &= 2\Gamma^2{}_{(12),1} + \Gamma^0{}_{(10),1} + 2\Gamma^2{}_{(12)}(\Gamma^2{}_{(12)} - \Gamma^1{}_{11}) \\
&\quad + \Gamma^0{}_{(01)}(\Gamma^0{}_{(01)} - \Gamma^1{}_{11}) + 2(\Gamma^2{}_{[13]})^2, \\
R_{22} &= -\Gamma^1{}_{22,1} + \Gamma^3{}_{(23),2} + (\Gamma^3{}_{(23)})^2 + 2\Gamma^1{}_{[23]}\Gamma^1{}_{[12]} \\
&\quad - \Gamma^1{}_{22}(\Gamma^1{}_{11} + \Gamma^0{}_{(01)}) = R_{33}\operatorname{cosec}^2\theta,
\end{aligned}
$$

$$R_{00} = -\Gamma^1{}_{00,1} - 2(\Gamma^3{}_{(02)} \sin^2 \theta)^2$$
$$+ \Gamma^1{}_{00}(\Gamma^0{}_{(01)} - \Gamma^1{}_{11} - 2\Gamma^2{}_{(12)}),$$

and

$$R_{[23]} = -\Gamma^1{}_{[23],1} + 2\Gamma^1{}_{22}\Gamma^2{}_{[13]} - \Gamma^1{}_{[23]}(\Gamma^1{}_{11} + \Gamma^0{}_{(01)}).$$

Hence the nontrivial solutions of the field equations

$$R_{(\mu\nu)} = 0 = R_{[[\mu\nu],\lambda]}$$

become

$$R_{00} = 0, \quad R_{11} = 0, \quad R_{22} = 0 \quad \text{and} \quad R_{[23]} = c\sin\theta, \tag{11}$$

where c is another real constant of integration (to be identified presently with the electric charge in electrostatic units). In terms of the variables defined in (5) it can be easily verified that these equations are equivalent to the following set of three second-order equations

$$P'' + \left(\frac{x'}{x} + \frac{y''}{y'}\right)P' + \frac{2}{x}(cu - \beta) = 0, \tag{12}$$

$$Q'' + \left(\frac{x'}{x} + \frac{y''}{y'}\right)Q' + \frac{2}{x}(u + c\beta) = 0, \tag{13}$$

$$2P'' - P'^2 + Q'^2 - 2\frac{y''}{y'}P' = 0, \tag{14}$$

together with

$$w^2 = (1/\lambda^2)y'^2 = k^2 y/x \tag{15}$$

where λ is another constant. We shall consider only the case

$$\lambda \neq 0, \tag{16}$$

that is when

$$y = \gamma U \neq \text{constant}.$$

Then, if we also write

$$c = \tan\varepsilon; \mu = 2\lambda^{-2}\sec\varepsilon; y = \exp\sigma; R = \sigma + P \text{ and } S = Q - \varepsilon, \tag{17}$$

and take σ as the independent variable, the field equations to be solved become:

$$\frac{d^2 S}{d\sigma^2} + \mu e^R \cos S = 0 = \frac{d^2 R}{d\sigma^2} - \mu e^R \sin S \tag{18}$$

and

$$2\frac{d^2 R}{d\sigma^2} = \left(\frac{dR}{d\sigma}\right)^2 - \left(\frac{dS}{d\sigma}\right)^2 - 1. \tag{19}$$

Equations (18) can be reduced to a single complex equation

$$\mathrm{d}^2 Z/\mathrm{d}\sigma^2 + \mu \exp(-iZ) = 0, \tag{20}$$

by the substitution

$$Z = S + iR. \tag{21}$$

The general solution of equation (20) is

$$\exp(iZ) = (v/2\zeta_0)\{1 - \cos(\sqrt{\zeta_0}(\sigma - \sigma_0))\}, \tag{22}$$

where $v = 2i\mu$ is pure imaginary and ζ_0 and σ_0 are complex constants of integration.

We obtain the real form of the Vanstone solution (published in 1962) on letting

$$\sqrt{\zeta_0} = p + iq, \quad \sigma_0\sqrt{\zeta_0} = s_1 + is_2 \tag{23}$$

with p, q, s_1 and s_2 real and

$$\left.\begin{array}{l} \lambda_1 = \dfrac{2pq\mu}{(p^2 + q^2)^2} = \mu, \sin\delta; \lambda_2 = \dfrac{\mu(p^2 - q^2)}{(p^2 + q^2)^2} = \mu, \cos\delta; \\[3mm] \chi_1 = p\sigma - s_1; \chi_2 = q\sigma - s_2; \\[2mm] \Phi = (1 - \cos\chi_1 \cosh\chi_2)/\sin\chi_1 \sinh\chi_2. \end{array}\right\} \tag{24}$$

Then

$$\exp(-R) = \frac{1}{y\rho} = \mu_1(\cosh\chi_2 - \cos\chi_1) \tag{25}$$

and

$$\tan S = \frac{\beta - cu}{u + c\beta} = \frac{\Phi - \tan\delta}{1 + \Phi\tan\delta},$$

from which we can calculate (using equation (15)) all the components of $g_{\mu\nu}$.

We must still satisfy equation (14). However, an elementary calculation shows that it becomes an identity if

$$p^2 = q^2 - 1, \tag{26}$$

so that we have only three independent arbitrary constants in the general solution.

It is important also to observe that the solution derived so far is expressed in terms of an arbitrary function y of the radial coordinate r. Hence, it is in fact an infinite set of possible static, spherically symmetric solutions. Moreover, appearance of three arbitrary constants of integration as well as an indefinite number of other constants in y makes interpretation of their physical meaning virtually impossible. On the other hand, the hypotheses of GFT require that the Vanstone solution should be compatible with the determination of the metric from the equation

$$a_{\mu\nu,\lambda} - \tilde{\Gamma}^{\sigma}_{(\mu\lambda)}a_{\sigma\nu} - \tilde{\Gamma}^{\sigma}_{(\lambda\nu)}a_{\mu\sigma} = 0 \tag{27}$$

(the metric hypothesis). I shall show in the next section that this requirement reduces Vanstone's set to just two possible expressions for y.

There is just one more remark which we can make before proceeding. It is that $a_{\mu\nu}$ defined by (27) is necessarily diagonal or form-invariant (universally spherically symmetric in the nomenclature of Gregory). Indeed, it can be easily checked that the only off-diagonal component the metric can have is

$$a_{01}$$

but that, unless a_{01} is zero, the field equations lead to a contradiction. The conclusion follows.

5.   The static, spherically symmetric metric of GFT

The Vanstone solution derived in the last section gives the static, spherically symmetric $g_{\mu\nu}$-field but not the corresponding geometry of the space–time. To determine the latter, we still have to solve equations (4.27). Only then may we be able to decide precisely to what physical situation the solution should correspond. Now, when $g_{\mu\nu}$ is given by (4.2) the symmetric part of the affine connection can be read off from the equations (4.7). It can be immediately verified that the components of the (diagonal) metric $a_{\mu\nu}$ are independent of t and φ, and that

$$a_{11} = a_0\alpha, \quad a_{22} = a_{33}\operatorname{cosec}^2\theta = b_0\rho, \quad a_{00} = y/y_0 \tag{1}$$

where a_0, b_0 and y_0 are nonzero constants. But the equations (4.27) also require that

$$\Gamma^0_{(10)}a_{00} + \Gamma^1_{00}a_{11} = 0 = \Gamma^2_{(12)}a_{22} + \Gamma^1_{22}a_{11} = \Gamma^2_{(03)}, \tag{2}$$

or, in the notation (4.5),

$$b_0\rho' + a_0(aQ' - BP') = 0 = y' + a_0 y_0 \gamma (\ln yU)', \tag{3}$$

$$wQ' = 0. \tag{4}$$

Hence the solution exists only if

$$\text{either } w = 0, \quad \text{or} \quad Q' = 0 \tag{5}$$

(or possibly both, though we shall soon find that the theory then reduces to GR: $u = w = 0$, $\tilde{\Gamma}^{\lambda}_{[\mu\nu]} = 0$). The second of the equations (3) can be integrated since

$$\gamma = y/u, \tag{6}$$

to give

$$y = \gamma U = y_1(1 + a_0 y_0/u), \tag{7}$$

where y_1 is yet another constant.

Let us consider first the case

$$Q' = 0.$$

If $u \neq 0$ then u is proportional to β while the first of the equations (3) gives

$$\beta = (b_0/a_0)\rho, \tag{8}$$

so that

$$\Gamma^1_{22} = -\beta\rho'/2\alpha\rho, \quad \Gamma^1_{[23]} = \tfrac{1}{2}k_0\rho'/\alpha, \tag{9}$$

where

$$k_0^2 = 1 - b_0^2/a_0^2$$

and

$$u = \pm k_0\rho, \quad k_0 \neq 0. \tag{10}$$

The field equations become

$$\left.\begin{aligned}
\left(\frac{\beta\rho'}{2\alpha\rho}\right)' + \tfrac{1}{2}(\ln \alpha y)'\frac{\beta\rho'}{2\alpha\rho} - 1 &= 0, \\[2mm]
\left(\frac{\rho'}{\alpha}\right)' + \tfrac{1}{2}(\ln \alpha y)'\rho'/\alpha + 2c/k_0 &= 0,
\end{aligned}\right\} \tag{11}$$

and

$$2P'' - P'^2 + P'(\ln (x/y))',$$

whence, from the last of these,

$$P'^2 x/\rho y = \kappa^2, \quad \text{a constant,} \tag{12}$$

and we also have

$$w^2 = k^2 y/x = (k^2/\lambda^2)y'^2, \tag{13}$$

(from the integral of $g^{[01]}{}_{,1} = 0$ and the definitions (4.5)). Hence

$$\alpha = \rho^2/x = \rho^2 y'^2/\lambda^2 y,$$

and therefore

$$y' = \pm (\lambda/\kappa)\rho'/\rho^{3/2},$$

which contradicts equation (7). Thus

$$\text{either } \lambda = 0 \quad \text{or} \quad u = 0. \tag{14}$$

In the first case y is a constant and therefore (from the second of equations (3), U and so) ρ is a constant which contradicts the first of equations (11). But it then follows that $Q' = 0$ implies

$$u = 0, \tag{15}$$

and since ρ can not vanish

$$k_0 = 0. \tag{16}$$

Hence, from the second of equations (11)

$$c = 0 \tag{17}$$

and this means that the field equations become 'strong', the case which we rejected because it did not lead to the correct equations of motion of a charged test particle.

Therefore a spherically symmetric, static solution of the field equations is possible only if

$$w = 0, \tag{18}$$

when

$$U = 1, \quad y = \gamma,$$

and

$$\Gamma^0_{(01)} = \tfrac{1}{2}\gamma'/\gamma, \quad \Gamma'_{00} \equiv \tfrac{1}{2}\gamma'/\alpha. \tag{19}$$

The components a_{11} and a_{22} are still given by (1), and so is a_{00} but it is more convenient to write instead

$$a_{00} = \gamma_0\gamma$$

(γ_0 replacing $y^{-1}{}_0$). The algebraic conditions (2) now become

$$\tfrac{1}{2}b_0\rho' + a_0\alpha\Gamma^1_{22} = 0 = \tfrac{1}{2}\gamma'(\gamma_0 + a_0) \tag{20}$$

so that

$$\Gamma^1_{22} = -(b_0/2a_0)\rho'/\alpha,$$

and either

$$\gamma = \text{constant},$$

or

$$a_0 = -\gamma_0.$$

It does not really matter which of the last two we adopt since even if we take γ nonconstant, coordinates will be found later in which the coefficient of $\mathrm{d}t^2$ in the metric relation

$$\mathrm{d}s^2 = a_{\mu\nu}\mathrm{d}x^\mu\mathrm{d}x^\nu, \quad x^0 = t, \tag{21}$$

is unity. Let us therefore write (without loss of generality)

$$a_0 = b_0 = -\gamma_0 = -1. \tag{22}$$

We then have

$$\sin Q = 1 - \frac{\rho}{\rho_0}, \quad \cos Q = \frac{u}{\rho}, \quad Q' = -\frac{\rho'}{\rho\sqrt{(2\rho_0/\rho - 1)}}, \\ \beta = \rho\left(1 - \frac{\rho}{\rho_0}\right), \quad \Gamma^1_{22} = -\frac{\rho'}{2\alpha}, \quad \text{and} \quad \Gamma^1_{[23]} = \frac{\rho}{2\alpha\sqrt{(2\rho_0/\rho - 1)}} \right\}$$

$$(23)$$

where ρ_0 is a constant. Since also $y = \gamma$, the field equations become

$$\left(\ln\frac{x}{y}\right)' = 2P' - \left(\ln\alpha\gamma\right)', \tag{24}$$

$$2P'' + P'^2 + Q'^2 - (\ln\alpha\gamma)'P' = 0, \tag{25}$$

$$\left(\frac{\rho'}{\alpha}\right)' + (\ln\alpha\gamma)'\frac{\rho'}{\alpha} - \frac{\rho'^2}{\alpha(2\rho_0 - \rho)} - 2 = 0, \tag{26}$$

and

$$\left(\frac{\rho'}{\alpha\sqrt{(2\rho_0/\rho - 1)}}\right) + \tfrac{1}{2}(\ln\alpha\gamma)'\frac{\rho'}{\alpha\sqrt{(2\rho_0/\rho - 1)}}$$

$$- \frac{\rho'^2}{\alpha\rho\sqrt{(2\rho_0/\rho - 1)}} + 2c = 0, \tag{27}$$

with

$$\gamma'^2 = \lambda^2\alpha\gamma/\rho^2. \tag{28}$$

Equation (24) gives at once

$$\alpha_y = \rho'^2/\mu\rho(2\rho_0 - \rho) \tag{29}$$

where μ is a nonzero constant of integration. If we also write

$$v\rho_0 = -\lambda/\sqrt{\mu}$$

we find, from equation (28), that

$$\alpha = \frac{\rho'^2}{\mu\rho(2\rho_0 - \rho)(\omega + v\sqrt{(2\rho_0/\rho - 1)})},$$

$$\gamma = \omega + v\sqrt{(2\rho_0/\rho - 1)}, \quad \omega = \text{constant}. \tag{30}$$

Moreover, it can be verified easily that the condition of compatibility of the remaining field equations is

$$\mu\omega\rho_0 - 2 = \mu v\rho_0 - 2c. \tag{31}$$

Of course, ρ remains an undetermined function of r. At first sight this seems to be exactly the case of the Vanstone solution but in fact is quite different. Vanstone found expressions for the components of the physical field when a given form of the solution would presumably correspond to some physical law. We have seen too that the symmetric part of the fundamental tensor can not represent the metric. On the other hand, the latter is precisely what is given by equations (30) and the metric determines the basic laws of measurement which have both a physical and a geometrical meaning. It is difficult to see how this could be valid without a definition of the radial coordinate which, as far as an observer is concerned is within his/her competence to define.

If then we require, as is natural, that

$$a_{22} = -\rho = -r^2, \tag{32}$$

and put

$$2\rho_0 = r_0^2, \tag{33}$$

the corresponding solution becomes

$$\left. \begin{aligned} \alpha &= \frac{1}{(1 - r^2/r_0^2)(1 + c\sqrt{(r_0^2/r^2 - 1)})}, \\ \gamma &= (1 + c\sqrt{(r_0^2/r^2 - 1)}), \\ u &= (2r^4/r_0^2)\sqrt{(r_0^2/r^2 - 1)}, \end{aligned} \right\} \tag{34}$$

since the constant ω can be clearly absorbed into $x^0 = t$. This solution is unique.

6.   The Coulomb law

Results of the last section give the general solution in the case of static spherical symmetry of the GFT field equations. The solution can be expected to correspond to the field of a stationary, electric point-charge/mass. We shall show now that such a charge gives rise to an inverse square, Coulomb field as well as an already obtained, definite geometry of the space–time in its vicinity.

I have identified previously (chapter 4, section 7) the electromagnetic intensity tensor

$$f_{\mu\nu} = R_{[\mu\nu]} \tag{1}$$

on the grounds that it yielded an acceptable form of the Lorentz force for the motion of a charged test particle and that it was exactly a Maxwell-like tensor (i.e. the curl of a 4-vector potential). This identification is now reinforced by the solution of the field equations.

Let us in fact recall the Papapetrou form (2.19) of the fundamental tensor. It follows from the equation (2.20) that a constant value of u and w vanishing, corresponds to a radial field component. The same is true of the skew part of the Ricci tensor (viz. 4.7.9). Hence equation (9), together with the field equation

$$R_{[23]} = c \sin \theta \tag{2}$$

implies immediately that the constant c represents a stationary, electric point-charge (strictly speaking, it is a non-dimensional charge ratio e/e_0). Moreover, the only component of a static spherically symmetric field is then in the radial direction:

$$c\hat{r}/r^2. \tag{3}$$

Therefore, the Coulomb law of electrostatic force is obeyed exactly. This is a firm prediction (albeit locally untestable since it is classical) of GFT, notwithstanding a correction to the law of motion demanded by the theory. On the other hand, the space–time in the vicinity of a stationary, point-mass charge is given by

$$ds^2 = (1 + c\sqrt{(r_0^2/r^2 - 1)})dt^2$$
$$- \frac{dr^2}{(1 - r^2/r_0^2)(1 + c\sqrt{(r_0^2/r^2 - 1)})} - r^2 d\Omega^2, \tag{4}$$

valid (since the components of a Riemannian metric tensor should be real) for

$$r < r_0. \tag{5}$$

If we return to the notation of the last section and write

$$\alpha\gamma = \frac{1}{1 - r^2/r_0^2}, \quad \gamma = 1 + c\sqrt{(r_0^2/r^2 - 1)},$$

the geodesic equations of the space (4), are

$$\left.\begin{array}{l} \ddot{t} + \dfrac{\gamma'}{\gamma}\dot{t}\dot{r} = 0, \\[2mm] \ddot{r} + \dfrac{1}{2}\dfrac{\alpha'}{\alpha}\dot{r}^2 + \dfrac{1}{2}\dfrac{\gamma'}{\alpha}\dot{t}^2 - \dfrac{\gamma}{\alpha}(\dot{\theta}^2 + \sin^2\theta\,\dot{\varphi}^2) = 0, \\[2mm] \ddot{\theta} + \dfrac{2}{r}\dot{r}\dot{\theta} - \sin\theta\cos\theta\,\dot{\varphi}^2 = 0, \\[2mm] \ddot{\varphi} + \dfrac{2}{r}\dot{r}\dot{\varphi} + 2\cot\theta\,\dot{\theta}\dot{\varphi} = 0, \end{array}\right\} \tag{6}$$

with dots denoting differentiation with respect to s, and dashes with respect to r.

Equations (6) represent the equations of motion of a neutral test particle in the gravitational field of the point-mass/charge c. Clearly, they allow a plane orbit

$$\theta = \tfrac{1}{2}\pi (d\theta/ds = 0 = d^2\theta/ds^2), \tag{7}$$

when

$$\gamma t = l, \quad r^2\dot{\varphi} = h, \tag{8}$$

l and h constant, and

$$1 = \frac{l^2}{\gamma} - \alpha\dot{r}^2 - \frac{h^2}{r^2}. \tag{9}$$

Clearly, we can put, without loss of generality,

$$l = 1.$$

Let us now suppose that

$$r \ll r_0 \tag{10}$$

and put, as usual,

$$r = 1/v. \tag{11}$$

Then, equations (8) and (9) give the differential equation of the orbit

$$\left(\frac{dv}{d\varphi}\right) = \frac{1}{r_0^2} + \left(\frac{2m}{h^2} - \frac{3m}{r_0^2}\right)v - v^2 + 2mv^3, \tag{12}$$

where we neglected powers of r/r_0^2 higher than the second (and also mr/r_0^2), and wrote

$$cr_0 = -2m. \tag{13}$$

To the above approximation therefore, the orbit is the same as in the Schwarzschild case in GR but with v displaced to

$$v - v_0$$

where v_0 is a real (positive) root of the cubic

$$\frac{1}{r_0^2} - \left(\frac{2m}{h^2} - \frac{3m}{r_0^2}\right)v_0 - v_0^2 - 2mv_0^3 = 0, \tag{14}$$

and the angular momentum h is replaced by

$$h(1 - 3h^2/2r_0^2)^{-1/2}. \tag{15}$$

It is nevertheless unlikely that this deviation from GR can be tested on a laboratory, or perhaps even 'astrophysical', scale. The space–time (4) appears to have a cut-off at

$$r = r_0, \tag{16}$$

and the impossibility of reconciling this with the electrostatic Coulomb law of force will provide us (in the next chapter) with an argument for introducing a cosmological interpretation of this result and a unique model of the universe as a prediction of GFT.

This is the most likely area in which an empirical verification of GFT may be sought, a conclusion which may be drawn also from the fact that locally the gravitational and electromagnetic fields decouple to a very high degree of accuracy. We already know that the spherically symmetric electrostatic field is strictly Coulomb. Furthermore, the metric (4) becomes

$$O(r/r_0),$$ (17)

$$ds^2 = (1 + cr_0/r)dt^2 - \frac{dr^2}{(1 + cr_0/r)} - r^2 d\Omega^2,$$ (18)

and so, with the notation (13), is simply the Schwarzschild metric, and even

$$O(r^2/r_0^2)$$

is hardly distinguishable empirically from the general relativistic case. Hence if tests on a laboratory (including astrophysical) scale are desired, they must depend either on hitherto unknown solutions or perhaps on GFT leading to electrodynamics of the Born–Infeld–Plebanski, nonlinear type. Of course, the metric (4) is quite different from the Reissner–Nordström GR solution occurring in the Einstein–Maxwell theory, but an enormous electrostatic charge would be required to produce empirically perceptible results.

In conclusion, let us, however, observe that since GFT allows only an electric point-charge it follows that no structureless (i.e. spherically symmetric) magnetic monopoles can exist. The concept of symmetry of course applies here in a strictly macroscopic sense although later (see chapter 7) we shall introduce into the nonsymmetric theory at least the idea of spinors and of a Dirac electrodynamics.

6 The cosmological model as a consequence of the generalised field theory

1. Local and cosmic coordinates

The unique, static, spherically symmetric solution (5.6.4) for the space–time metric about a structureless, massive electric charge breaks down at a radial distance

$$r = r_0$$

from the origin of the coordinate system where the charge is supposed to be situated. However, this 'singularity' can be easily removed as follows. If we write

$$r' = \frac{r_0}{\sqrt{(r_0^2/r^2 - 1)}}, \tag{1}$$

and take r' as the new 'radial' coordinate while keeping t, θ and φ the same, the metric becomes

$$ds^2 = (1 + cr_0/r')dt^2 - \frac{r_0^4 dr'^2}{(r_0^2 + r'^2)^2(1 + cr_0/r')} - \frac{r_0^2 r'^2}{r_0^2 + r'^2}d\Omega^2. \tag{2}$$

Dropping the dash on r (I shall refer to it simply as the new coordinate until a more physically meaningful name for it is presently found), and adopting the notation (5.6.13), the metric relation is

$$ds^2 = \gamma dt^2 - w^2 dr^2/\gamma - wr^2 d\Omega^2, \tag{3}$$

where

$$\gamma = 1 - 2m/r, \tag{4}$$

is the Schwarzschild factor and

$$w = r_0^2/(r_0^2 + r^2). \tag{5}$$

The cosmological considerations of this chapter depend on a comparison of the metrics (3) and (5.6.4). The latter represents the geometry around a stationary (with respect to an observer), spherically symmetric, 'point'-charge identified in the field equations with the dimensionless constant c. It is thus a charge ratio e/e_0, its lack of dimensions being explicitly illustrated by the intervention of the constant $2m$ in the formula (5.6.13). This enables us to regard the constant $2m$ as a characteristic mass, something perhaps like the 'mass of distant stars'-effect in Mach's principle. When r_0 is allowed to tend to infinity in the metric relation (3), the latter becomes as already noted in section 5.6 just the Schwarzschild metric but, if $2m$ is to remain finite, the charge must then vanish. For the present solution, this charge is the only source of a nongravitational, that is an electromagnetic, field. The field is then also removed and we revert to general relativity in its primitive form.

Hence GFT requires r_0 to be finite as a precondition of presence of an electromagnetic interaction. It must be observed also, that removal of the singularity at $r = r_0$ (in the old coordinates) is, like all such singularity shifting processes in Riemannian geometry, still embodied in the coordinate transformation formula (1).

This means that, from a physical point of view, the new and the old coordinates are radically different and refer to different situations. And there is something very curious about the old system. According to the identification of the electromagnetic field of the last chapter, a static point-charge gives rise to a classical Coulomb field. The space–time metric, however, has a cut-off at a finite distance from the charge. How can this be since there is no question that the old coordinates represent the local situation? Electromagnetic fields with a cut-off have been contemplated but not with an exact, inverse square law of force demanded by our theory in the static conditions.

The only way out of the resulting paradox is not merely to regard r_0 as large (and presumably c as necessarily 'small') but to render meaningless the otherwise natural question of what happens for (old)

$$r > r_0. \tag{6}$$

This will be the case, of course, if we take r_0 as the radius of a (static, spherically symmetric model of a) universe.

The new coordinates (from which the $r = r_0$ singularity has been removed) will then become cosmological coordinates and we can, and will from now on, refer to the two systems as 'local' and 'cosmological' respectively.

As far as an observer is concerned, it is up to him/her to choose what should be measured as a 'radial' distance at least within a wide latitude of mathematical freedom. When a choice is made, it is then possible for purely conceptual reasons, whatever physical sense this may make, to let the selected coordinate tend to infinity. If now the cosmological radial coordinate

$$r \to \infty \tag{7}$$

in formula (3), the latter collapses into

$$ds^2 = dt^2 - r_0^2(d\theta^2 + \sin^2\theta\, d\varphi^2) \tag{8}$$

whose spatial part is an Euclidean 2-sphere of radius r_0. Its surface can not be crossed by a signal sent by the observer and can be regarded as the outermost boundary of the universe. The fact that it is not flat will have considerable repercussions on the definition of cosmological concepts. We may also note that for a stationary, with respect to our observer, source of light

$$ds = \sqrt{(1 - 2m/r)}dt \approx (1 - m/r)dt, \tag{9}$$

reinforcing our previous conclusion that GFT is unlikely to be locally testable.

2. The light tracks (null geodesics)

The inevitable conclusion of the last section is that the static, spherically symmetric solution of the (physical) field equations plus geometry leads in the generalised field theory to the model of the universe given by the metric relation (1.3)

$$ds^2 = (1 - 2m/r)dt^2 - \frac{r_0^2 dr^2}{(r_0^2 + r^2)^2(1 - 2m/r)} - \frac{r_0^2 r^2}{r_0^2 + r^2}d\Omega^2$$

Comparison of observational evidence with whatever consequences may be drawn from the above alone can determine whether this resembles the overall features of the world in which we live. It is therefore important for GFT to investigate physical meaning of the above geometry.

The appearance in the above relation of a Schwarzschild-like singularity at

$$r = 2m \tag{1}$$

seems to indicate that r should not be interpreted as the radial distance from an observer, although it is about this surface that the world is symmetric. On the other hand, there is no reason, at present, to doubt that t can be regarded as time.

In general relativistic physics, cosmological information is obtained along the null geodesics of a Riemannian space–time which represent the light tracks of incoming signals. That this must be the case also in GFT is born out by the local non-interaction of gravitational and electromagnetic fields demanded by the weak principle of geometrisation on which the theory is founded.

Let us therefore start by determining the differential equations of the null geodesics in the r–t system. At any rate, the limiting form of the metric (as $r \to \infty$) is independent of the coordinate choice.

If dots denote derivatives with respect to the parameter s, we easily find that

$$\ddot{t} + \frac{2m}{r(r-2m)}\dot{r}\dot{t} = 0,$$

$$\ddot{r} - \left(\frac{2r}{r_0^2 + r^2} + \frac{m}{r(r-2m)}\right)\dot{r}^2 - (r-2m)$$
$$\times(\dot{\theta}^2 + \sin^2\theta\,\dot{\varphi}^2) + \frac{m}{r_0^4 r^3}(r-2m)(r_0^2 + r^2)\dot{t}^2 = 0, \tag{2}$$

$$\ddot{\theta} + \frac{2r_0^2}{r(r_0^2 + r^2)}\dot{r}\dot{\theta} - \sin\theta\cos\theta\,\dot{\varphi}^2 = 0,$$

$$\ddot{\varphi} + \frac{2r_0^2}{r(r_0^2 + r^2)}\dot{r}\dot{\varphi} + 2\cot\theta\,\dot{\theta}\dot{\varphi} = 0.$$

Hence, as before, we can put

$$\theta = 0, \quad \theta = \tfrac{1}{2}\pi, \tag{3}$$

for coplanar paths, when

$$\frac{\dot{\varphi}r_0^2 r^2}{r_0^2 + r^2} = h, \quad \dot{t} = \frac{kr}{r-2m}, \tag{4}$$

where h and k are constants. With dashes denoting derivatives with respect to φ, we now get, putting $v = 1/r$,

$$v'^2 + v^2 = \frac{k^2 - 1}{h^2} - \frac{1}{r_0^2} + 2m\left(\frac{1}{h^2} + \frac{1}{r_0^2}\right)v + 2mv^3, \tag{5}$$

or

$$v'' + v = m(h^2 + r_0^2)/h^2 r_0^2 + 3mv^2. \tag{6}$$

The null geodesics are obtained as usual by letting

$$h \to \infty, \tag{7}$$

in the form

$$v'' + v = m/r_0^2 + 3mv^2. \tag{8}$$

We shall find later that our model resembles in many respects a de Sitter universe although it is not empty. It follows also from equation (8) that its light tracks are not, as in the latter, straight lines. Of more interest, however, are radial null geodesics because, as already noted, it is along these that the observer derives his/her knowledge of the universe. And we can easily convince ourselves that a radial geodesic remains radial with respect to the observer, as an immediate consequence of the fact that the expression

$$(1 - 2m/r)\frac{dt^2}{d\lambda^2} - \frac{r_0^4}{(r_0^2 + r^2)^2(1 - 2m/r)}\left(\frac{dr}{d\lambda}\right)^2$$

$$- \frac{r_0^2 r^2}{r_0^2 + r^2}\left(\left(\frac{d\theta}{d\lambda}\right)^2 + \sin^2\theta\left(\frac{d\varphi}{d\lambda}\right)^2\right) = 0 \tag{9}$$

remains an integral of the geodesic equations whatever initial conditions may be imposed on it.

Thus, putting

$$\dot\varphi = 0 \text{ as well as } \theta = \tfrac{1}{2}\pi \text{ and } \dot\theta = 0, \tag{10}$$

we get

$$0 = dt^2 - \frac{r_0^4 dr^2}{(r_0^2 + r^2)^2(1 - 2m/r)}, \tag{11}$$

or

$$t = \frac{r_0^2}{r_0^2 + 4m^2}\left(m\ln\frac{(r - 2m)^2}{r_0^2 + r^2} + r_0\tan^{-1}\frac{r}{r_0}\right) + \text{constant.} \tag{12}$$

Consider now (cf. Eddington 1924, section 70), light signals or pulses emitted by an atom at (t, r). The time of arrival of a single pulse emitted at $t = t_0$ at some point P at (still in the t, r coordinate system) where an observer (who may be considered to remain at rest) is situated is given by

$$t = t_0 + \frac{r_0^2}{r_0^2 + 4m^2}\left(m\ln\frac{(r - 2m)^2}{r_0^2 + r^2} + r_0\tan^{-1}\frac{r}{r_0}\right) + \text{constant,} \tag{13}$$

if we assume that the pulse travelled along the observer's line of sight. Hence for successive pulses

$$\Delta t = \Delta t_0 + \frac{r_0^2 r}{(r - 2m)(r_0^2 + r^2)}\Delta r$$

$$= \left(1 + \frac{r_0^2 r}{(r - 2m)(r_0^2 + r^2)}\frac{\Delta r}{\Delta t_0}\right)\frac{\Delta t_0}{\Delta s}\Delta s, \tag{14}$$

where, if the square of velocity of the atom is neglected,

$$\Delta s = \sqrt{((r - 2m)/r)\Delta t_0}. \tag{15}$$

Thus, the ratio of durations between two pulses emitted by a distant atom to that of a similar atom at $P(\Delta t_P)$ is

$$\frac{\Delta t}{\Delta t_P} = \sqrt{\left\{\frac{r(r_P - 2m)}{r_P(r - 2m)}\right\}} + \sqrt{\left(\frac{r_P - 2m}{r_P}\right)\frac{r_0^2}{r_0^2 + r^2}\left(\frac{r}{r - 2m}\right)^{3/2}}\frac{\Delta r}{\Delta t_0}$$

(16)

where the velocity dependent second term on the right hand side of equation (16) represents the red shift (to the first order in the velocity).

3. Isotropy and expansion of the world

One of the gravest difficulties of cosmological theory is that most of the observational data are theory dependent, that is their meaning can be interpreted only by assuming a particular theoretical explanation. It has always been a cherished astronomical rule that terrestrially discovered laws can and must be extrapolated in an attempt to understand evidence. It is of no concern to us that physical laws might be different elsewhere in the universe as long as we can interpret what we see in terms of the laws we know. An example of the application of this rule is given by the acceptance of extragalactic recession: red shift of the light from distant galaxies is a Doppler effect because every alternative explanation contradicts some apparently well-understood law of physics. It rests, however, on the assumptions that atoms behave in the same way everywhere in the universe. The assumption and the rule are an expression of the only way in which we can discuss without conceptual chaos what goes on where we can not reach. All the same, the implication is that we must always carefully scrutinise the technique by which given conclusions are reached.

As far as a theory is concerned, it is clear that evidence whose meaningful interpretation involves its assumption can not be used for its verification. Most of the current cosmologies employ general relativity in one way or another, but the trouble with GR is that very few of its predictions are locally verifiable. Indeed, they still virtually reduce to the three original 'crucial tests': the motion of the perihelion, bending of light and the gravitational redshift (a very different thing from the Hubble effect!), derived from the Schwarzschild solution. Since there is matter in the universe (a tautological assertion on which all physics depends), cosmological theory requires field equations with a nonvanishing energy–momentum tensor. By what I have said, however, its conclusions can hardly be regarded as an empirical confirmation (or otherwise) of, say, the principle of equivalence. This is why it is vitally important that the foundations from which a cosmological model is obtained should be as sound as possible, both logically and empirically, and in particular that they should be free from assumptions which may not be warranted. There appear to be two 'facts' which, with all the above reservations, seem to be sufficiently well established and uncontroversial. These are the overall isotropy of the universe and the Hubble law of expansion.

A model which denies these is surely suspect from the start. Now, we can easily cast the metric (1.3) into the isotropic form

$$ds^2 = f^2(\rho)dt^2 - g^2(\rho)(d\rho^2 + \rho^2 d\Omega^2). \tag{1}$$

In fact, we require

$$g d\rho = \frac{r_0^2 dr}{(r_0^2 + r^2)\sqrt{(1 - 2m/r)}} \quad \text{and} \quad g\rho = \frac{rr_0}{\sqrt{(r_0^2 + r^2)}},$$

$$f^2(\rho) = 1 - \frac{2m}{r}, \tag{2}$$

$$\frac{d\rho}{\rho} = \frac{r_0 dr}{\sqrt{\{(r_0^2 + r^2)(r^2 - 2mr)\}}}.$$

A substitution

$$\frac{r}{2m} = \frac{\zeta + \frac{2}{3}r_0^2}{\zeta - \frac{1}{3}r_0^2 - 4m^2/r_0}$$

gives

$$\rho = \exp[-2^{2/3}\sqrt{r_0}\,\wp^{-1}(\zeta, 2^{2/3}(\tfrac{1}{3}r_0^2 - 4m^2),$$
$$- (2r_0/27)(r_0^2 + 36m^2))], \tag{3}$$

where $\wp$ is the Weierstrass elliptic function. Since we can now determine ρ as an invertible function of r, we can also find f and g from equations (2).

As a matter of interest we can also transform the metric into its Szekeres–Kruskal form

$$ds^2 = \frac{dy^2 - dv^2}{g(r)} - \frac{r_0^2 r^2}{r_0^2 + r^2} d\Omega^2, \tag{4}$$

where

$$v = aX \cosh(t/a), \quad u = aX \sinh(t/a),$$
$$a = \lambda r_0/(1 + \lambda^2), \quad \lambda = 2m/r_0,$$
$$X^2 = (1 - \lambda \cot(z/r_0))g(z),$$
$$g = \frac{1 - (1 + \lambda^2)\cos^2(z/r_0)}{1 + \lambda \cot(z/r_0)} \exp\left(\frac{2z}{\lambda r_0}\right),$$

and
$$r = r_0 \tan(z/r_0).$$
However, $g(r)$ vanishes when $r = 2m$ so that the above transformation does not shift the singular surface as in the case of the Schwarzschild geometry.

This result is rather important because it emphasises the physical significance of the constant $2m$. I have so far refrained from speculating as to what it might be, except for its vague connection with charge which has somehow disappeared in cosmological considerations (I shall say something about this in the last chapter). Some light on its possible meaning and value will be thrown by the Hubble expansion formula.

Before we turn to the latter, let us introduce a yet another transformation putting simply
$$\rho = \sqrt{(1 - 2m/r)}, \tag{5}$$
when
$$ds^2 = \rho^2 dt^2 - \frac{16m^2 d\rho^2}{((1-\rho^2)^2 + \lambda^2)^2} - \frac{4m^2 d\Omega^2}{(1-\rho^2)^2 + \lambda^2}. \tag{6}$$
ρ apparently can now go from 0 to ∞, but for
$$r < 2m,$$
it reverses roles with t. If we then put
$$t = R, \quad \rho = iT,$$
the metric becomes
$$ds^2 = \frac{16m^2 dT^2}{((1+T^2)^2 + \lambda^2)^2} - T^2 dR^2 - \frac{4m^2 d\Omega^2}{(1+T^2)^2 + \lambda^2}. \tag{7}$$
For a radial, null geodesic, we get
$$\frac{4mT\dot{T}}{(1+T^2)^2 + \lambda^2} = k, \quad \text{a constant,} \tag{8}$$
with an approximate red-shift formula
$$\frac{\Delta T - \Delta \tau}{\Delta \tau} \sim \frac{k[(1+T^2)^2 + \lambda^2]}{4mT} - 1. \tag{9}$$
The all important expansion law will be obtained by suppressing the angular dependence in (1.3):
$$ds^2 = \left(1 - \frac{2m}{r}\right)dt^2 - \frac{r_0^4 dr^2}{(r_0^2 + r^2)^2(1 - 2m/r)}$$

whence, with dots denoting differentiation with respect to s,

$$\ddot{r} + \frac{m}{r_0^2}\left(1 + \frac{r^2}{r_0^2}\right)^2 - \frac{2r\dot{r}^2}{r_0^2}\left(1 + \frac{r^2}{r_0^2}\right)^{-1} = 0. \qquad (10)$$

This equation can be easily integrated once, to give

$$\dot{r}^2 = \left(k + \frac{2m}{r}\right)\left(1 + \frac{r^2}{r_0^2}\right)^2, \qquad (11)$$

where k is an arbitrary constant which, however, appears to have to be positive for real velocities. Then substituting back into equation (10) we obtain

$$\ddot{r} + \frac{m}{r^2}\left(1 + \frac{r^2}{r_0^2}\right)^2 - \frac{2m}{r_0^2}\left(k + \frac{2m}{r}\right)\left(1 + \frac{r^2}{r_0^2}\right) = 0. \qquad (12)$$

Now it is well known that, for the de Sitter world, the corresponding result is exactly (Hubble law)

$$\ddot{r} = r/r_0^2. \qquad (13)$$

When

$$2m \ll r \ll r_0, \qquad (14)$$

we obtain from the equation (12), and to a high degree of accuracy,

$$\ddot{r} = 2kr/r_0^2. \qquad (15)$$

Hence, Hubble's expansion law holds in the GFT model but only as an approximation, for relatively nearby galaxies. It must be noted that the law is stated as a second-order relation (rather than the more usual $v = r/H$) because in the GFT model we can not approach the origin $r = 0$ without complications. The galaxies for which the law holds are therefore those for which we put (with tongue in cheek!)

$$\left(1 + \frac{r^2}{r_0^2}\right)\sqrt{\left(k + \frac{2m}{r}\right)} \sim \sqrt{\left(k^2 + \frac{2kr^2}{r_0^2}\right)}.$$

This is not bad if

$$m/r \ll r^2/r_0^2. \qquad (16)$$

'The galaxies must be near an observer but sufficiently far from whatever $r = 2m$ may mean.'

It now looks as though the critical distance $2m$ should have something to do with the early universe. Presumably the universe would never have exploded if gravitational attraction overwhelmed all other forces although it can, so to speak, always be thought of as a 'black hole' since by definition light can not escape from it. We can think of $2m$ as the radius of the primeval atom, not indeed a singularity which surely is abhorred by Nature no less than a vacuum (and whatever some theoreticians may speculate using hardly applicable quantum concepts), but an object with a large enough density though not too large for macroconcepts to fail. Then, taking density of matter in our universe as perhaps $10^{-28} - 10^{-31}$ g cm^{-3}, and r_0 as the distance of a Hubble horizon from the observer, the inequality (14), the region where the observed expansion holds, can be something of the order of

$$10^{15}\,\text{cm} \ll r \ll 10^{30}\,\text{cm}. \tag{17}$$

4. An oscillating universe?

In the de Sitter cosmology the radius r_0 is identified with the inverse H^{-1} of the Hubble constant. The GFT model requires r_0 to be the radius of the universe but then the Hubble law contains also the constant k which we must now try to estimate. To this end we shall write

$$\dot{r}^2 = \dot{t}^2 (dr/dt)^2, \tag{1}$$

and, eliminating $\dot{t}^2$ from the metric relation, find that

$$\frac{dr}{dt} = \pm \sqrt{\left(\frac{k + 2m/r}{k+1}\right)\left(1 - \frac{2m}{r}\right)\left(1 + \frac{r^2}{r_0^2}\right)}. \tag{2}$$

Let us now take the positive sign (expansion) and suppose that speed of light $(dr/dt = 1)$ is reached when

$$r = r_0 \gg 2m. \tag{3}$$

Then

$$k = \tfrac{1}{3} \tag{4}$$

and the approximate Hubble law (3.15) becomes

$$\ddot{r} = 2r/3r_0^2 = H^2 r \tag{5}$$

The GFT relation between the Hubble constant and the radius of the universe is therefore

$$H^{-1} = r_0 \sqrt{\tfrac{3}{2}}, \tag{6}$$

a slight correction on the standard interpretation.

Applying now the inequality (3.16) to equation (2), we get approximately (though the further we go from $r = 2m$ the better the approximation)

$$dr/dt = \tfrac{1}{2}(1 + \tfrac{3}{2}H^2 r^2) \tag{7}$$

whence, if we also take

$$r = 2m \text{ when } t = 0, \tag{8}$$

$$r = \left| \frac{2m + (\sqrt{6}/3H)\tan(\sqrt{6}Ht/4)}{1 - mH\sqrt{6}\tan(\sqrt{6}Ht/4)} \right|. \tag{9}$$

This represents an oscillating universe. For an observer, a receding galaxy reaches infinity (coordinate and not physical) after a time t from the initial explosion given by

$$mH\tan(\sqrt{6}Ht/4) = 1/\sqrt{6},$$

or approximately, when

$$t \sim 2\pi/H\sqrt{6} \sim 2/H.$$

Therefore, granting the approximations, the universe appears to oscillate with a period $4/H$.

5. Geometry at infinity

We have noted in the first section of this chapter that at r-infinity the metric collapses into (1.8)

$$ds^2 = dt^2 - r_0^2(d\theta^2 + \sin^2\theta \, d\varphi^2).$$

Curiously, this is not a space of constant curvature, though of course its 'spatial' section is. This can be verified at once since the only significant component of its Riemann–Christoffel tensor is

$$R_{2332} = \sin^2\theta, \tag{1}$$

and does not satisfy the constant curvature condition

$$R_{\lambda\mu\nu\kappa} = K(g_{\lambda\nu}g_{\mu\kappa} - g_{\mu\nu}g_{\lambda\kappa}), \tag{2}$$

for the metric diag $(1, -1, -\sin^2\theta)$ when, for example $\lambda = \nu = 1$, $\mu = \kappa = 2$.

It may be intuitively obvious that time in the space (1.8) is absolute but we shall now record a general proof of this result. We must solve the Killing equations

$$\xi_{\lambda;\mu} + \xi_{\mu;\lambda} = 0, \tag{3}$$

which, if we write

$$(x^1, x^2, x^3) = (t, \theta, \varphi), \tag{4}$$

become

$$\left. \begin{array}{l} \xi_{1,1} = 0, \quad \xi_{1,2} + \xi_{2,1} = 0, \quad \xi_{3,1} + \xi_{1,3} = 0, \quad \xi_{2,2} = 0, \\ \xi_{2,3} + \xi_{3,2} - 2\cot\theta\,\xi_3 = 0, \quad \xi_{3,3} + \sin\theta\cos\theta\,\xi_2 = 0. \end{array} \right\} \tag{5}$$

It follows easily that

$$\xi_1 = -f\theta + g, \quad \xi_2 = ft + h, \quad \xi_3 = (f'\theta - g')t + w(\theta, \varphi), \tag{6}$$

where f, g and h are functions on φ only, w is a function of θ, φ, as shown, and dashes denote derivatives with respect to φ.

Substituting this solution into the last two of the Killing equations (5), we obtain

$$f't = h' + f't + \partial w/\partial \theta - 2wt\theta[(f'\theta - g')t + w] = 0, \tag{7}$$

$$(f''\theta - g'')t + \partial w/\partial \varphi + \sin \theta \cos \theta (ft + h) = 0. \tag{8}$$

It follows that

$$f = 0, \quad g = g_0, \text{ a constant,}$$

and

$$\partial w/\partial \theta - 2 \cot \theta w + h' = 0 = \partial w/\partial \varphi + h \sin \theta \cos \theta. \tag{9}$$

The condition

$$\partial^2 w/\partial \varphi \partial \theta = \partial^2 w/\partial \theta \partial \varphi$$

of integrability of the equations (9) gives

$$h'' + h = 0,$$

or

$$h = h_0 \cos (\varphi + \varepsilon), \tag{10}$$

h_0, ε constant. Then, k_0 being another constant,

$$w = -h_0 \sin (\varphi + \varepsilon) \sin \theta \cos \theta + k_0 \sin^2 \theta, \tag{11}$$

and the general solution of the equations is easily seen to be a linear combination of the four covariant Killing vectors

$$\left.\begin{array}{l} {}^1\xi_\lambda = (1, 0, 0), \quad {}^2\xi_\lambda = (0, \cos \varphi, -\sin \varphi \sin \theta \cos \theta), \\ {}^3\xi_\lambda = (0, -\sin \varphi, -\cos \varphi \sin \theta \cos \theta), \quad {}^4\xi_\lambda = (0, 0, -\sin^2 \theta) \end{array}\right\} \tag{12}$$

If we raise λ with the help of the metric (1.8), the infinitesimal generators of the complete group of motions G_4 in our space–time:

$$X_\alpha = {}^\alpha\xi^\lambda \partial/\partial x^\lambda, \tag{13}$$

become

$$\left.\begin{array}{l} X_1 = \dfrac{\partial}{\partial t}, \quad X_2 = -\cos \varphi \dfrac{\partial}{\partial \theta} + \sin \varphi \cot \theta \dfrac{\partial}{\partial \varphi}, \\[2mm] X_3 = \sin \varphi \dfrac{\partial}{\partial \theta} + \cos \varphi \cot \theta \dfrac{\partial}{\partial \varphi}, \quad X_4 = \dfrac{\partial}{\partial \varphi}. \end{array}\right\} \tag{14}$$

Hence they satisfy the commutation relations

$$\left.\begin{array}{l} [X_1, X_a] = 0, \quad a = 1, 2, 3, \\ [X_2, X_3] = X_4, \quad [X_3, X_4] = X_2, \quad [X_4, X_2] = X_3. \end{array}\right\} \tag{15}$$

In other words, G_4 splits into a translation G, in t, absolute up to a choice of origin, and a Euclidean rotation O_3. This proves our intuitive assertion about time at infinity.

6. Time and distance

The central thesis of generalised field theory is that the distribution of matter represented by an energy–momentum tensor $T_{\mu\nu}$ is to be calculated after the field equations have been solved under some enabling and *a priori* symmetry restrictions. A coordinate system is presupposed in selecting the latter but this does not mean that it is necessarily correctly chosen. The matter tensor is itself independent of this choice, its components, however, are not, and it is the components of $T_{\mu\nu}$ that we habitually identify with physically measurable quantities. Already Eddington (1924, section 77) has emphasised the distinction between invariant and relative mass, which (or rather its associated mass density) is of critical importance to cosmology; its value determines in the GR models whether the universe is closed or open. Whenever we talk about mass density therefore, there arises the crucial problem whether we mean an abstract invariant or a seemingly measurable but coordinate dependent, relative density. In GFT, the problem is further complicated by the world model not being flat at infinity. However confident we may feel about the choice of radial distance from an observer which appears to be arbitrary, it is the time parameter for which we can find criteria of 'correctness'.

Let us first transform the GFT cosmological metric (1.3) into a system in which the coefficient of dt^2 is unity. This form is, of course, useful also in comparing the GFT universe with the constant curvature models derived from a Robertson–Walker metric. If

$$\xi = t + h(r), \quad \tau = t + g(r), \tag{1}$$

where

$$\frac{dh}{dr} = \pm \frac{w}{\gamma \sqrt{(1-\gamma)}}, \quad \frac{dg}{dr} = \pm \frac{w\sqrt{(1-\gamma)}}{\gamma}, \tag{2}$$

we easily find that the relation (1.3) becomes

$$ds^2 = d\tau^2 - R^2 d\xi^2 - P^2 d\Omega^2, \tag{3}$$

where

$$R^2 = \frac{2m}{r} = 1 - \gamma, \quad P^2 = wr^2 = \frac{4m^2}{\lambda^2 + R^4}, \quad \lambda = \frac{2m}{r_0}. \tag{4}$$

If we put

$$T = \tau - \xi, \tag{5}$$

equations (2) give

$$\frac{dT}{dr} = \pm \frac{r_0^2}{\sqrt{(2m)r_0^2 + r^2}} \frac{\sqrt{r}}{}, \tag{6}$$

or

$$\frac{dR}{dT} = \pm \frac{\lambda^2 + R^4}{4m}. \tag{7}$$

The choice of sign in the above equation is important but will be left till later, although we can see now that the metric components R and P have become functions of T only. The GFT universe is certainly not of the Robertson–Walker type.

Let us now consider the Killing equations. It is convenient to use the system

$$(x^0, x^1, x^2, x^3) = (T, \xi, \theta, \varphi), \tag{8}$$

in which the metric relation is

$$ds^2 = dT^2 + 2dTd\xi + Q^2 d\xi^2 - P^2 d\Omega^2, \tag{9}$$

$$Q^2 = \gamma = 1 - R^2.$$

With dashes denoting derivatives with respect to T, the Christoffel brackets are

$$\begin{Bmatrix} 0 \\ 01 \end{Bmatrix} = -\frac{R'}{R}, \quad \begin{Bmatrix} 0 \\ 11 \end{Bmatrix} = -\frac{Q^2 R'}{R},$$

$$\begin{Bmatrix} 0 \\ 22 \end{Bmatrix} = \begin{Bmatrix} 0 \\ 33 \end{Bmatrix} \mathrm{cosec}^2\,\theta = -\frac{Q^2 PP'}{R^2},$$

$$\begin{Bmatrix} 1 \\ 01 \end{Bmatrix} = \frac{R'}{R}, \quad \begin{Bmatrix} 1 \\ 11 \end{Bmatrix} = \frac{R'}{R}, \quad \begin{Bmatrix} 1 \\ 22 \end{Bmatrix} = \begin{Bmatrix} 1 \\ 33 \end{Bmatrix} \mathrm{cosec}^2\,\theta = \frac{PP'}{R^2},$$

$$\begin{Bmatrix} 2 \\ 02 \end{Bmatrix} = \frac{P'}{P}, \quad \begin{Bmatrix} 2 \\ 33 \end{Bmatrix} = -\sin\theta\cos\theta,$$

$$\begin{Bmatrix} 3 \\ 03 \end{Bmatrix} = \frac{P'}{P}; \quad \begin{Bmatrix} 3 \\ 23 \end{Bmatrix} = \cot\theta, \tag{10}$$

and the Killing equations become

$$X_{0,0} = 0, \quad X_{0,1} + X_{1,0} + 2(R'/R)(X_0 - X_1) = 0,$$

$$X_{0,2} + X_{2,0} - (2P'/P)X_2 = 0 = X_{0,3} + X_{3,0} - \frac{2P'}{P}X_3,$$

$$X_{1,1} + (R'/R)(Q^2 X_0 - X_1) = 0,$$

$$X_{1,2} + X_{2,1} = 0 = X_{1,3} + X_{3,1},$$

$$X_{2,2} + (PP'/R^2)(Q^2 X_0 - X_1) = 0 = X_{2,3} + X_{3,2} - 2\cot\theta X_3,$$

$$X_{3,3} + \sin\theta\cos\theta X_2 + (PP'/R^2)(Q^2 X_0 - X_1) = 0, \tag{11}$$

where

$$(X_0, X_1, X_2, X_3)$$

are the components of a Killing vector X_μ in the coordinate system (8). A straightforward inspection of the equations (11) shows that their only separable solution is the time-like vector

$$X_\mu = (1, Q^2, 0, 0). \tag{12}$$

This is mapped into

$$X_\mu = (Q^2, 0, 0, 0), \tag{13}$$

by the transformation

$$t = \int \frac{dT}{Q^2} + \xi, \quad r = \xi, \quad \theta = \theta, \quad \varphi = \varphi. \tag{14}$$

We must note that the coordinates t and r are different from in (1.3), but $Q^2 = \gamma$ is, because of (14), a function of

$$\rho = t - r,$$

only. In the new coordinates, the metric relation is

$$\begin{aligned} ds^2 &= Q^4 dt^2 - Q^2 R^2 (dr^2 - 2dt\,dr) - P^2 d\Omega^2 \\ &= Q^2 dt^2 - Q^2 R^2 d\rho^2 - P^2 d\Omega^2. \end{aligned} \tag{15}$$

The transformation from the coordinate system (8) to

$$(t, \rho, \theta, \varphi)$$

is given by

$$\xi = t - \rho, \quad T = \int Q^2 d\rho, \quad \theta = \theta, \quad \varphi = \varphi, \tag{16}$$

Q (as well as R and P) being a function of ρ only.

Because of the solution (13), the parameter t now appears to be the correct choice for the time. Let us, however, recall the all important fact that cosmological evidence, that is information which we receive from distant sources, arrives *ex hypothese* along radial null geodesics only. And whether we consider the metric (3) or the metric (15) as the correct expression, radial null geodesics are given by

$$d\xi/d\tau = -1/R \quad \text{and} \quad d\rho/dt = -1/R, \tag{17}$$

the choice of sign being dictated by light having been emitted in the past as far as an observer is concerned. In the first of equations (17), R is a function of $T = \tau - \xi$ given by (subject to arbitrary initial conditions) the equation (7), and in the second, of ρ. It is of course the same function of T in both equations and

$$d\rho/dR = \pm 4m/(1 - R^2)(\lambda^2 + R^4). \tag{18}$$

But in such considerations as redshift it is the velocity, $d\xi/d\tau$ or $d\rho/dt$, that is measured empirically. It follows that it is difficult to determine on empirical grounds which of the systems (τ, ξ) and (t, ρ) is the proper determinant of the fundamental measures of time and of radial distance of a source from the observer. Are there any other criteria by which the choice might be made?

7. Distribution of matter in the universe

Let us consider the distribution of matter in the GFT universe. Since (in the original coordinates)

$$2m < r < \infty, \tag{1}$$

$Q^2 = \gamma = 1 - 2m/r$ varies from 0 to 1 (and R^2 from 1 to 0). As in chapter 2, the energy–momentum tensor is calculated from

$$T_{\mu\nu} = -\frac{1}{\kappa}G_{\mu\nu} = -\frac{1}{\kappa}(R_{\mu\nu} - \frac{1}{2}g_{\mu\nu}(R - 2\Lambda)), \tag{2}$$

the geometrical tensors being constructed from the Christoffel brackets, and Λ being the cosmological constant (required by the conservation property of $T_{\mu\nu}$ and not by an extraneous hypothesis). Also, by analogy with general relativity if for no other reason, we can put

$$\kappa = 8\pi N/c^2 \tag{3}$$

when the speed of light in vacuum is not unity and N is the Newtonian gravitational constant. Strictly speaking there is no justification in GFT for the relation (3) except that in the absence of nongravitational fields, say for a neutral fluid or in the case of gravitational waves, equation (2) could still be expected to yield the correct geometry corresponding to a given (electrically) neutral source. GFT refers to fundamental fields only!

For the same reason we can write heuristically

$$T_{\mu\nu} = (\rho_0 + p)u_\mu u_\nu - g_{\mu\nu}p, \tag{4}$$

with ρ_0 denoting mass density and p the pressure (u_μ being the covariant velocity vector). The GFT results should at any rate approximate to this form however unwarranted (prerelativistic) it may be in the strict sense. But then we immediately see (for slowly moving matter) that Λ can not be zero since this would correspond to negative pressure everywhere. The most likely values of Λ are in fact

$$\Lambda_1 = 1/r_0^2, \quad \Lambda_2 = 2/r_0^2 \quad \text{or} \quad \text{the de Sitter } \Lambda_3 = 3/r_0^2 \tag{5}$$

We can draw up table (6) in the three coordinate systems we have used, for slow matter.

$$T_{00} = \rho_0 = -\frac{1}{\kappa}G_{00} \qquad T_{11} = -g_{11}\,p = -\frac{1}{\kappa}G_{11} \qquad T_{22} = -g_{22}\,p = -\frac{1}{\kappa}G_{22}$$

	$-\frac{1}{\kappa}G_{00}$	$-\frac{1}{\kappa}G_{11}$	$-\frac{1}{\kappa}G_{22}$
$(T,\xi,\theta,\varphi) \equiv S_1$	$\dfrac{1}{\kappa}\left(\dfrac{\gamma+2}{r_0^2} - \Lambda\right)$	$\dfrac{\gamma}{\kappa}\left(\dfrac{3\gamma}{r_0^2} - \Lambda\right)$	$\dfrac{p^2}{\kappa}\left(\Lambda - \dfrac{\gamma}{r_0^2}\right)$
$(\tau,\xi,\theta,\varphi) \equiv S_2$	$\dfrac{1}{\kappa}\left(\dfrac{\gamma+2}{r_0^2} - \Lambda\right)$	$\dfrac{1-\gamma}{\kappa}\left(\Lambda + \dfrac{2-3\gamma}{r_0^2}\right)$	$\dfrac{p^2}{\kappa}\left(\Lambda - \dfrac{\gamma}{r_0^2}\right)$
$(t,\rho,\theta,\varphi) \equiv S_3$	$\dfrac{\gamma}{\kappa}\left(\dfrac{3\gamma}{r_0^2} - \Lambda\right)$	$\gamma\dfrac{(1-\gamma)}{\kappa}\left(\Lambda - \dfrac{\gamma}{r_0^2}\right)$	$\dfrac{p^2}{\kappa}\left(\Lambda - \dfrac{\gamma}{r_0^2}\right)$

$$(6)$$

The most instructive are the extreme cases when

$$\gamma = 0 \text{ and } \gamma = 1, \tag{7}$$

and clearly we shall derive most information by examining the T_{00} component of the energy–momentum tensor. Thus, at $r = 2m$ (in the original coordinates, $\gamma = 0$), T_{00} is as in table (8)

	Λ_1	Λ_2	Λ_3
S_1	$1/\kappa r_0^2$	0	$-1/\kappa r_0^2$
S_2	$1/\kappa r_0^2$	0	$-1/\kappa r_0^2$
S_3	0	0	0

$$(8)$$

and at infinity ($\gamma = 1$) T_{00} is as in table (9).

	Λ_1	Λ_2	Λ_3
S_1	$2/\kappa r_0^2$	$1/\kappa r_0^2$	0
S_2	$2/\kappa r_0^2$	$1/\kappa r_0^2$	0
S_3	$2/\kappa r_0^2$	$1/\kappa r_0^2$	0

$$(9)$$

When Λ takes on the de Sitter value $3/r_0^2$ ($\Lambda = \Lambda_3$), the negative value of mass density on (but also close to) the critical surface $r = 2m$ is difficult to understand in the empirically possible coordinate system $(\tau, \xi, \theta, \varphi)$.

Indeed, tables (8) and (9) suggest that we have

$$\Lambda = 2/r_0^2 \tag{10}$$

when the fact that density vanishes on $r = 2m$ can be interpreted as meaning that 'all of the primeval fire-ball blew up'.

With this value of the cosmological constant the covariant components of the Einstein tensor in the coordinate system $(\tau, \xi, \theta, \varphi)$ are finally

$$G_{00} = -\frac{\gamma}{r_0^2}, \quad G_{11} = -(1-\gamma)\left(\frac{4-3\gamma}{r_0^2}\right),$$

$$G_{22} = G_{33}\operatorname{cosec}^2\theta = \frac{\lambda^2(\gamma - 2)}{\lambda^2 + (1-\gamma)^2}. \tag{11}$$

The coordinate system (S_2) is particularly convenient, not only because of its Robertson–Walker form, but also because in it,

$$G_{00} = G^0{}_0, \tag{12}$$

and this saves the uncertainty, which necessarily arises in GFT, of which of these components should be associated with the (relative) mass density. We thus take equations (11) as determining the distribution of matter in the universe.

If we assume that

$$g^{\mu\nu}u_\mu u_\nu = 1, \tag{13}$$

the above distribution may be regarded as a perfect fluid of relative mass density

$$\rho_0 = (3\gamma - 2)/\kappa r_0^2, \tag{14}$$

moving with four velocity

$$u_\mu = \pm(1/\sqrt{\gamma}, \, -(1-\gamma)/\sqrt{\gamma}, \, 0, \, 0) \tag{15}$$

under the pressure

$$P = (2 - \gamma)/\kappa r_0^2. \tag{16}$$

However, this interpretation breaks down close to $r = 2m$, and it is perhaps better to regard the matter distribution (11) as something other than a perfect fluid.

8.   Volume and mass of the world

We can readily calculate the relative volume of a space-like section and the relative mass of the world represented by our model. The former, that is a 3-volume of the universe, is given by

$$V = 4\pi \int_{\xi_0}^{\xi_1} Rp^2 \mathrm{d}\xi, \tag{1}$$

on the section

$$\tau = \text{constant.} \tag{2}$$

In formula (1), ξ_0 corresponds to $R = 1$ and ξ_1 to $R = 0$. Since on account of (2),

$$d\xi = -dT \tag{3}$$

we must take the positive sign in (6.7):

$$dR/dT = +(\lambda^2 + R^4)/4m, \tag{4}$$

for V to be positive. Then

$$V = 4\pi r_0^3 \left(\tan^{-1} \frac{1}{\lambda} + \frac{\lambda}{1+\lambda^2} \right) \approx 4\pi r_0^3 \left(\tfrac{1}{2}\pi + \frac{\lambda}{1+\lambda^2} \right) \tag{5}$$

for a small λ and hence is $O(\lambda)$, the same as the corresponding volume of an Einstein universe.

Similarly, taking

$$T_{00} = \rho = \frac{1}{\kappa} \left(\frac{2+\gamma}{r_0^2} - \Lambda \right) = \frac{1}{\kappa} \left(\frac{3}{r_0^2} - \Lambda - \frac{R^2}{r_0^2} \right), \tag{6}$$

the total (relative) mass

$$\begin{aligned} M &= \frac{64\pi m^3}{\kappa c^2} \int_0^1 \frac{(3/r_0^2 - \Lambda - x/r_0^2)dx}{(\lambda^2 + x^2)^2} \\ &= \frac{4\pi r_0^3}{\kappa c^2} \left(\frac{3}{r_0^2} - \Lambda \right) \left(\tan^{-1} \frac{1}{\lambda} + \frac{\lambda}{1+\lambda^2} \right) - \frac{4\pi r_0 \lambda^3}{\kappa(1+\lambda^2)}. \end{aligned} \tag{7}$$

This result shows better than anything else that we can not have the de Sitter cosmological constant

$$\Lambda = 3/r_0^2.$$

When

$$\Lambda = 2/r_0^2$$

we get, for small λ,

$$M = 2\pi^2 r_0/c^2 \kappa. \tag{8}$$

The main value of these results, however, is theoretical, namely the fixing of the sign in the equations (6.7) and the rejection of the de Sitter cosmological constant.

9. Cosmic observables

I want to reiterate once more the facts that must be kept in mind whenever discussing the universe at large, that our information arrives along radial, null geodesics and that its very understanding is all too often theory dependent. I have been concerned so far with the overall structure of the GFT cosmological model. We have seen that it admits the two features of which we can be least uncertain from the empirical point of view, namely isotropy and expansion. But what are the actual observables on which cosmological speculation is based? In this section, I shall briefly sketch some consequences of the now unambiguous model (at any rate, we have settled the sign of (6.7)), which may lead to its observational verification.

Although τ or t represents the correct time parameter it is T which is the argument of the metric coefficients. T has the appearance of a retarded time and is the quantity which is presumably measured by an observer's clock. Let us start with the $R-T$ equation

$$\mathrm{d}T/\mathrm{d}R = 4m/(\lambda^2 + R^4) = P^2/m, \tag{1}$$

and define the red shift z by the standard formula

$$z = R_\mathrm{a}/R - 1, \tag{2}$$

where the suffix 'a' denotes the value of R at the arrival time of a signal, that is at the point where an observation is made. Then

$$\mathrm{d}z/\mathrm{d}T = -\{\lambda^2(1+z)^4 + R_\mathrm{a}^4\}/4mR_\mathrm{a}(1+z)^2, \tag{3}$$

or, up to z^3

$$[R_a^3(1+v^2)/4m\}\Delta T = z + (1 - 2\mu^2)z^2 + \tfrac{1}{3}(1 - 14\mu^2 + 16\mu^4)z^3 + O(z^4), \tag{4}$$

where

$$v = \lambda/R_\mathrm{a}^2, \quad \mu^2 = v^2/(1+v^2),$$

and

$$\Delta T = T_\mathrm{a} - T.$$

Let us now write

$$R_\mathrm{a}^3(1+v^2)/4m\Delta T = \delta T \tag{5}$$

which is equivalent to choosing units of time so that the Hubble constant is unity. Inverting the series (4), we find that

$$z = \delta T[1 + (2\mu^2 - 1)\delta T + \tfrac{1}{3}(5 + 2\mu^2 + 8\mu^4)\delta T^2] + \ldots \tag{6}$$

It now follows that the deceleration parameter

$$q = -R\ddot{R}/\dot{R}^2 = -p^2R^4/m^2, \tag{7}$$

and is necessarily negative. This result, however, can not be now excluded, as in the standard theories, on the grounds that it must lead to a negative energy density. It does not, because the latter is independently calculated in GFT (section 7). At the observer

$$q_a = -4/(1+v^2) \tag{8}$$

and hence, is not only negative but also rather large. This is an inevitable consequence of the present theory.

For an incoming signal, the track is given by

$$d\xi/d\tau = -1/R.$$ (9)

Since for most of the universe

$$R \gg \sqrt{(2m/r_0)} = \sqrt{\lambda},$$

let us write

$$r = \sqrt{\lambda}/R$$

when, for our new r,

$$r \ll 1,$$

and

$$\frac{dT}{dr} = -\frac{4}{\lambda^{3/2}} \frac{r^2}{1 + r^4}$$ (10)

and, to a high degree of accuracy,

$$R \approx \left(\frac{4m}{3}\right)^{1/3} (T_a - T)^{-1/3}.$$ (11)

We can finally consider cosmological distance when determined by the apparent size of an observed object. Let us assume that its extremities are at the same radial distance ξ_0, from the observer, given by integrating equation (9). Then the diameter of the object, or the local separation of its extremities, say l, is given by

$$l = Pd\theta,$$ (12)

so that in terms of the red shift, its angular diameter is

$$\frac{l}{P} = \frac{l}{2m} \sqrt{\left(\lambda^2 + \frac{R_a^4}{(1 + z)^4}\right)}$$ (13)

and as z varies from zero to infinity, the angular diameter varies between the limits $(l/2m)\sqrt{(\lambda^2 + R_a^4)}$ and l/r_0 respectively.

On the other hand, the distance can be obtained from

$$d\xi/d\tau = d\xi/dT = -1/R,$$

or

$$d\xi/dz = 4m(1 + z)^3/R_a^4 + \lambda^2(1 + z)^4.$$ (14)

Hence

$$\xi = \frac{m}{\lambda^2} \ln \frac{R_a^4 + \lambda^2(1 + z)^4}{\lambda^2 + R_a^4}.$$ (15)

It follows that there are no horizons in any finite part of the universe which is occupied by matter or fields.

For small red shifts

$$\xi = \frac{4m}{\lambda^2 + R_{\mathrm{a}}^4} z \left(1 + \frac{3R_{\mathrm{a}}^4 - \lambda^2}{2(R_{\mathrm{a}}^4 + \lambda^2)} z \right) + O(z^3) \tag{16}$$

Analogous formulae involving luminosity distance can be obtained as in the standard cosmologies.

We must note in conclusion that for small z, that is close to an observer, formula (13) must be replaced by l/ξ which becomes infinite at $z = 0$ as required conceptually.

10. Inside 2m

The region (with the original r)

$$r < 2m \tag{1}$$

must be excluded from the observable universe, and has been in the above calculations. Although the r–T equation becomes singular only at $r = 0$ (and $r = \infty$), the roles of space and of time become reversed. Writing in fact,

$$r = \tau, \quad t = x,$$

we have, for $\tau < 2m$,

$$ds^2 = \frac{w^2}{2m/\tau - 1} d\tau^2 - (2m/\tau - 1)dx^2 - w\tau^2 d\Omega^2$$

$$= dT^2 - F^2(T)dx^2 - G^2(T)d\Omega^2 \tag{2}$$

where

$$T = \int_0^\tau \frac{w\sqrt{\tau}\,d\tau}{\sqrt{(2m - \tau)}}, \quad w = \frac{r_0^2}{r_0^2 + \tau^2}, \quad F^2 = \frac{2m}{\tau} - 1,$$

$$G^2 = \frac{r_0^2 \tau^2}{r_0^2 + \tau^2}$$

A standard calculation now shows that whatever positive Λ we choose, the energy density turns out to be negative. Hence, no matter how we interpret this region, the laws of physics therein must be very different from what we are accustomed to. Could this possibly provide us with a mechanism for the initial (and presumably recurring) explosion or the 'Big Bang'?

11. Cosmology without a cosmological principle

We must now consider a remarkable aspect of the GFT cosmology. With the possible exception of the de Sitter model, which is a direct solution of the empty field equations of GR with a vanishing integration constant (m, but of course with a cosmological constant Λ), relativistic cosmologies rely on some form of a cosmological principle. The latter is usually a smoothing-out hypothesis imposed *a priori* on the distribution of matter in the universe. Examples are provided by the Friedmann principle leading to the Robertson–Walker metrics, the 'perfect' principle of steady state cosmologies or even the Big Bang hypothesis on which most of the speculations on the early conditions of the world are based. There is, however, another way of looking at the cosmological principles. It is that they are the criterion by which this or that solution of the field equations is asserted to be a model of the universe. In this sense, then, all cosmologies involve a cosmological principle.

Nevertheless, the above is not the usual sense of the term and when the criterion of choice is other than a postulate about the state of matter, it is legitimate to refer to the model as a cosmology without a cosmological principle. The GFT cosmology and the de Sitter world are perhaps the only two examples of such a theory.

The distribution of matter in the generalised field theory is determined after the field equations have been solved under a hypothetical symmetry restriction. Identification of the individual components of an energy–momentum tensor with physically meaningful quantities must still depend on the assumption of a given form for it, but the components themselves are well-defined functions of the space–time coordinates.

We may recall that the foregoing cosmological model arose from our interpretation of the spherically symmetric static local solution. More exactly, the reason for regarding it as a cosmology was enshrined in the difficulties which a common-sense concept of electromagnetism would otherwise encounter. Is this sufficient for proposing a cosmology? I think so, provided it conforms to at least what seems to be known about the universe, and because the nature of the electromagnetic field appears to be well understood locally. It is then significant that GFT predicts the structure of the world without invoking extraneous principles. It is perhaps even more significant that it does this uniquely. If spherical symmetry is agreed upon, there is no room in it for the multitudinous models of GR.

12. Charge and mass

The Riemannian metric of GFT (1.6) has been interpreted as a cosmological relation outside the critical surface

$$r = 2m = -cr_0. \tag{1}$$

It defines there the fundamental laws of measurement in the presence of gravitational and electromagnetic fields which permeate the universe. If electromagnetic field is absent ($c = 0, r_0 = \infty$), the metric reduces to a Schwarzschild space–time which presumably would be the correct exterior geometry if the primeval fire-ball did not blow up. Hence the effect of electromagnetism manifests itself on a cosmological scale and, if the proposed model should prove to account well for the observed properties of the world, the existence of a cosmic electromagnetic field has to be inferred whatever mechanism (e.g. charge imbalance) may be argued for its origin. It is then a consequence of the assumed laws of physics.

The question still remains as to what is the relation between electric charge and mass. The effect mentioned previously (interpretation of the constant $2m$ as mass) could be called a 'Mach effect of the charge' but it would be more accurate to say that GFT requires the charge to manifest itself as mass on a cosmic scale. We conclude our discussion of cosmology with this tentative remark.

7 Generalised field theory and microphysics

1. Spinors in the nonsymmetric theory

In directing the development of GFT towards cosmological considerations, I have virtually abandoned hope that it can lead to significant results in the realm of quantum mechanics. Einstein felt that a comprehensive theory should describe the structure of all physical fields, macroscopic as well as microscopic; a conviction which stemmed from his being at odds with the principle of uncertainty underlying microphysical thinking. Contrary to this, we have seen that the success of GFT consists in its being regarded as a completion of general relativity rather than a search for something it can not contain. It is true that hitherto undiscovered solutions might describe more physics than the primitive fields we have found, but it is clearly unlikely that mere symmetries of space–time on which these solutions depend should be sufficient to account for all the diverse empirical features of elementary particles: spin, strangeness, colour, flavour, or what have you. The distinction between macrophysics and microphysics seems to lie at a more fundamental level of the concept of what constitutes an observation and an observable. In macrophysics, we practically disregard considerations of the process of observation. Field quantities, the solutions of the basic equations, are by definition meaningful and measurable. Indeed, macrophysics can be defined as that aspect of reality in which the interference of an observer, though not his/her interpretative faculty, can be neglected. And I have stressed throughout this monograph the macrophysical nature of the generalised field theory. The most that we can expect from it is to tell us where its meeting point with microphysics may be. One extension which is admissible because of the theory's Riemannian background is to a spinor formalism. If the latter can be regarded as belonging to quantum mechanics then this is one of the areas where the theories converge, although in the absence of at least the exchange relations we can not claim to have a quantum theory. It is both curious and satisfying, however, that the GFT outlined in the preceding chapters gives a quantum-related electrodynamics in a way which is more convincing than was possible in GR.

Throughout this chapter, I shall denote spinor indices by Latin capitals and assume that these take the values 1 and 2 only. Thus we have spinors

$$\varphi_A, \quad \psi^A \text{ and their complex conjugates } \varphi_{\dot{A}}, \quad \psi^{\dot{A}}, \tag{1}$$

etc. A spinor of rank 2 (a 2×2 matrix with complex elements) will be said to be Hermitian if

$$h_{\dot{A}B} = h_{B\dot{A}} \tag{2}$$

In this case, the diagonal elements are real and the off-diagonal ones are complex conjugates of each other, so that the spinor consists of four functions (of the space–time coordinates), just as a vector. There exist therefore mixed (vector–spinor) quantities σ (which can be represented in terms of Pauli spin matrices: see e.g. the excellent article by Bade & Jehle (1953) whose basic notation I have adopted) which relate spinors to vectors:

$$h_{\dot{A}B} = \sigma^\mu{}_{\dot{A}B} h_\mu, \quad h^\mu = \sigma^{\mu \dot{A}B} h_\mu. \tag{3}$$

Spinor indices will be raised or lowered with the help of a skewsymmetric, fundamental, metric spinor in any of its forms

$$\gamma_{AB}, \quad \gamma^{AB}; \quad \gamma_{\dot{A}\dot{B}}, \quad \gamma^{\dot{A}\dot{B}}. \tag{4}$$

We shall find presently that GFT requires γ to be a constant matrix

$$\gamma = \begin{bmatrix} 0 & 1 \\ -1 & 0 \end{bmatrix} = \bar{\gamma}. \tag{5}$$

We have

$$\varphi_A \psi^A = \varphi_A \gamma^{AB} \psi_B = -\gamma^{BA} \varphi_A \psi_B = -\varphi^B \psi_B \tag{6}$$

(so that $\varphi_A \varphi^A = 0$). We also have from (3) that the matrices are Hermitian with respect to their spinor indices. The (arbitrary) identification of the metric forms

$$a^{\mu\nu} h_\mu h_\nu \quad \text{and} \quad \gamma^{\dot{A}\dot{B}} \gamma^{CD} h_{\dot{A}C} h_{\dot{B}D} \tag{7}$$

yields the defining equations of spinor algebra:

$$\sigma^{\mu \dot{A}B} \sigma^\nu{}_{\dot{A}B} = a^{\mu\nu}, \tag{8}$$

$$\sigma^\mu{}_{\dot{A}B} \sigma^{\dot{C}D}{}_\mu = \delta^{\dot{C}}{}_{\dot{A}} \delta^D{}_B, \tag{9}$$

and

$$\sigma^\mu{}_{\dot{A}B} \sigma^{\nu \dot{A}C} + \sigma^\nu{}_{\dot{A}B} \sigma^{\mu \dot{A}C} = a^{\mu\nu} \delta^C{}_B. \tag{10}$$

We should note that it is the Riemannian metric tensor $a_{\mu\nu}$ and not, for example, the tensor $g_{(\mu\nu)}$ which necessarily appears in the above relations. Spinor algebra belongs to the background Riemannian space–time and not, at this stage, to the nonsymmetric 'physics'. This of course, is not surprising since spinors are a representation of the Lorentz group.

2. Spinor analysis

Let us define covariant derivatives of a contravariant spinor ψ^A by

$$D_\lambda \psi^A = \partial_\lambda \psi^A + \psi^B \left\{ \begin{matrix} A \\ \lambda B \end{matrix} \right\}, \quad \nabla_\lambda \psi^A = \partial_\lambda \psi^A + \psi^B \Gamma^A{}_{\lambda B},$$

$$\tilde{\nabla}_\lambda \psi^A = \partial_\lambda \psi^A + \psi^B \tilde{\Gamma}^A{}_{\lambda B}, \tag{1}$$

where the spin connections $\left\{ \begin{matrix} A \\ \lambda B \end{matrix} \right\}$, $\Gamma^A{}_{\lambda B}$, $\tilde{\Gamma}^A{}_{\lambda B}$ will be related to the

Christoffel brackets $\left\{ \begin{matrix} \lambda \\ \mu v \end{matrix} \right\}_a$ and the connections $\Gamma^\lambda{}_{\mu v}$, $\tilde{\Gamma}^\lambda{}_{\mu v}$ respectively and the operator ∂_λ replaces the comma notation of previous chapters. We shall refrain from attempting to use quantities such as $\Gamma^A{}_{B\lambda}$ since the spinor and tensor indices are basically distinct and as far as its effect on tensors is concerned ∇_λ will be invariably used for $\nabla^+{}_\lambda$. If the quantity

$$\varphi_A \psi^B$$

is regarded as a scalar, then

$$D_\lambda \Psi_A = \partial_\lambda \Psi_A - \Psi_B \left\{ \begin{matrix} B \\ \lambda A \end{matrix} \right\}, \quad \text{etc.} \tag{2}$$

Similarly, if for the purposes of differentiation, $\sigma^\lambda{}_{\dot{A}B}$ is treated as an outer product $h^\lambda \psi_{\dot{A}} \varphi_B$ (for the conjugates we, of course, have $\left\{ \begin{matrix} \dot{A} \\ \lambda \dot{B} \end{matrix} \right\}$ etc.), it follows from equation (1.8) that

$$D_\lambda \sigma^\mu{}_{\dot{A}B} = D_\lambda \sigma^{\mu \dot{A}B} = 0. \tag{3}$$

Let us also define the tensors $W_{\lambda \mu v}$, $\tilde{W}_{\lambda \mu v}$ symmetric in the second pair of indices ($W_{\lambda \mu v} = W_{\lambda v \mu}$) by

$$\nabla_\lambda a_{\mu v} = W_{\lambda \mu v}, \quad \tilde{\nabla}_\lambda a_{\mu v} = \tilde{W}_{\lambda \mu v}. \tag{4}$$

Tensor indices are still raised and lowered with the help of the metric $a_{\mu v}$.

We now observe that $\nabla_\lambda \sigma^\mu{}_{\dot{A}B}$ must be in general a nonvanishing linear combination of the σ vector–spinors since these span the spinor space. Thus

$$\nabla_\lambda \sigma^\mu{}_{\dot{A}B} = \partial_\lambda \sigma^\mu{}_{\dot{A}B} + \Gamma^\mu{}_{\nu\lambda}\sigma^\nu{}_{\dot{A}B} - \Gamma^{\dot{C}}{}_{\lambda\dot{A}}\sigma^\mu{}_{\dot{C}B} - \Gamma^C{}_{\lambda B}\sigma^\mu{}_{\dot{A}C}$$
$$= H^\mu{}_{\lambda\nu}\sigma^\nu{}_{\dot{A}B}, \tag{5}$$

where $H^\mu{}_{\lambda\nu}$ is a tensor. It follows from equations (4) and (1.8) that

$$H^\mu{}_{\lambda\rho}\sigma^\rho{}_{\dot{A}B}\sigma^{\nu\dot{A}B} + H^\nu{}_{\lambda\rho}\sigma^\mu{}_{\dot{A}B}\sigma^{\rho\dot{A}B} = -W^{\mu\nu}{}_\lambda,$$

or

$$H^{(\mu\nu)}{}_\lambda = -\tfrac{1}{2}W^{\mu\nu}{}_\lambda \tag{6}$$

and if we write also

$$\tilde{\nabla}_\lambda \sigma^\mu{}_{\dot{A}B} = \tilde{H}^\mu{}_{\lambda\nu}\sigma^\nu{}_{\dot{A}B}, \tag{7}$$

where $\tilde{H}^\mu{}_{\lambda\nu}$ is a new tensor, then

$$\tilde{H}^{(\mu\nu)}{}_\lambda = -\tfrac{1}{2}\tilde{W}^{\mu\nu}{}_\lambda. \tag{8}$$

As far as the fundamental spinor is concerned, we must have (Bade & Jehle 1953)

$$D_\lambda(\gamma_{\dot{A}\dot{B}}\gamma_{CD}) = \nabla_\lambda(\gamma_{\dot{A}\dot{B}}\gamma_{CD}) = \tilde{\nabla}_\lambda(\gamma_{\dot{A}\dot{B}}\gamma_{CD}) = 0. \tag{9}$$

Consider now the equation

$$\partial_\lambda \sigma^{\mu A\dot{B}} + \begin{Bmatrix} \mu \\ \rho\lambda \end{Bmatrix}\sigma^{\rho A\dot{B}} + \begin{Bmatrix} A \\ \lambda C \end{Bmatrix}\sigma^{\mu C\dot{B}} + \begin{Bmatrix} \dot{B} \\ \lambda\dot{C} \end{Bmatrix}\sigma^{\mu A\dot{C}} = 0. \tag{10}$$

Multiplying it by $\sigma^\nu{}_{\dot{B}D}$, adding the equation resulting from interchanging the vector indices μ and ν and recalling (1.10), we get

$$a_{\mu\nu}\begin{Bmatrix} \mu\nu & A \\ \lambda & B \end{Bmatrix} + 4\begin{Bmatrix} A \\ \lambda B \end{Bmatrix} + 2\begin{Bmatrix} \dot{C} \\ \lambda\dot{C} \end{Bmatrix}\delta^A{}_B = 0, \tag{11}$$

where

$$\begin{Bmatrix} \mu\nu & A \\ \lambda & B \end{Bmatrix} = 2(\partial_\lambda \sigma^{(\mu|A\dot{C}|})\sigma^{\nu)}{}_{\dot{C}B} + \begin{Bmatrix} \mu \\ \rho\lambda \end{Bmatrix}\sigma^{\rho AC}\sigma^\nu{}_{\dot{C}B}$$

$$+ \begin{Bmatrix} \nu \\ \rho\lambda \end{Bmatrix}\sigma^{\rho A\dot{C}}\sigma^\mu{}_{\dot{C}\dot{B}}. \tag{12}$$

Hence, when $A \neq B$, the sixteen real functions which comprise $\begin{Bmatrix} A \\ \lambda B \end{Bmatrix}$ are given by

$$\begin{Bmatrix} A \\ \lambda B \end{Bmatrix} = -\tfrac{1}{4}a_{\mu\nu}\begin{Bmatrix} \mu\nu & A \\ \lambda & B \end{Bmatrix}. \tag{13}$$

On the other hand, denoting the diagonal elements by

$$u_\lambda = \left\{ \begin{matrix} 1 \\ \lambda 1 \end{matrix} \right\}, \quad v_\lambda = \left\{ \begin{matrix} 2 \\ \lambda 2 \end{matrix} \right\}, \quad U_\lambda = a_{\mu\nu} \left\{ \begin{matrix} \mu\nu \\ \lambda \end{matrix} \middle| \begin{matrix} 1 \\ 1 \end{matrix} \right\},$$

$$V_\lambda = a_{\mu\nu} \left\{ \begin{matrix} \mu\nu \\ \lambda \end{matrix} \middle| \begin{matrix} 2 \\ 2 \end{matrix} \right\}, \tag{14}$$

we have

$$2u_\lambda + \bar{u}_\lambda + \bar{v}_\lambda + \tfrac{1}{2}U_\lambda = 0 = 2v_\lambda + \bar{v}_\lambda + \bar{u}_\lambda + \tfrac{1}{2}V_\lambda. \tag{15}$$

Hence, if we also write

$$R_\lambda = -\tfrac{1}{2}(U_\lambda + V_\lambda) \quad \text{and} \quad V_\lambda - U_\lambda = G_\lambda + iH_\lambda, \tag{16}$$

where G_λ and H_λ are real, equations (15) give the partial solution

$$\mathrm{Re}\, u_\lambda = \tfrac{1}{8}(R_\lambda + G_\lambda), \quad \mathrm{Re}\, v_\lambda = \tfrac{1}{8}(R_\lambda - G_\lambda),$$

$$\mathrm{Im}\, u_\lambda - \mathrm{Im}\, v_\lambda = \tfrac{1}{4}H_\lambda, \tag{17}$$

but

$$\mathrm{Im}\, u_\lambda + \mathrm{Im}\, v_\lambda \propto \text{an arbitrary vector.} \tag{18}$$

In the original analysis of van der Waerden & Infeld (see Bade & Jehle 1953) this vector was identified with an electromagnetic vector potential but in GFT we have already identified Γ_λ as the vector potential.

If we now assume the fundamental spinor γ_{AB} to be a constant matrix, the equation

$$D_\lambda \gamma_{AB} = 0 \tag{19}$$

will imply that (no summation)

$$\left\{ \begin{matrix} A \\ \lambda A \end{matrix} \right\} = 0. \tag{20}$$

It then follows from equations (5) and (7) that

$$\Gamma^C_{\ \lambda B} = \left\{ \begin{matrix} C \\ \lambda B \end{matrix} \right\} + \tfrac{1}{2}\sigma_\mu^{\ \dot{A}C}\sigma^\nu_{\ \dot{A}B}(N^\mu_{\ \nu\lambda} - H^\mu_{\ \lambda\nu}) - \tfrac{1}{2}\Gamma^{\dot{A}}_{\ \lambda\dot{A}}\delta^C_{\ B}, \tag{21}$$

and

$$\tilde{\Gamma}^C_{\ \lambda B} = \left\{ \begin{matrix} C \\ \lambda B \end{matrix} \right\} + \tfrac{1}{2}\sigma_\mu^{\ \dot{A}C}\sigma^\nu_{\ AB}(\tilde{N}^\mu_{\ \nu\lambda} - \tilde{H}^\mu_{\ \lambda\nu}) - \tfrac{1}{2}\tilde{\Gamma}^{\dot{A}}_{\ \lambda\dot{A}}\delta^C_{\ B}, \tag{22}$$

where

$$N^\lambda_{\ \mu\nu} = \Gamma^\lambda_{\ \mu\nu} - \left\{ \begin{matrix} \lambda \\ \mu\nu \end{matrix} \right\}_a, \quad \tilde{N}^\lambda_{\ \mu\nu} = \tilde{\Gamma}^\lambda_{\ \mu\nu} - \left\{ \begin{matrix} \lambda \\ \mu\nu \end{matrix} \right\}_a. \tag{23}$$

3. Some relations between 'tilded' and untilded tensors

The definition of the Schrödinger affine connection

$$\tilde{\Gamma}^{\lambda}{}_{\mu\nu} = \Gamma^{\lambda}{}_{\mu\nu} + \tfrac{2}{3}\delta^{\lambda}{}_{\mu}\Gamma_{\nu}$$

enables us to derive a number of relations between tilded and untilded quantities which will be required in the sequel. Thus, we immediately obtain from equations (2.4)

$$\tilde{W}_{\lambda\mu\nu} = W_{\lambda\mu\nu} - \tfrac{4}{3}\Gamma_{\lambda}a_{\mu\nu}. \tag{1}$$

It follows from (2.6) and (2.8) that

$$\tilde{H}_{\lambda(\mu\nu)} = H_{\lambda(\mu\nu)} + \tfrac{2}{3}\Gamma_{\lambda}a_{\mu\nu}. \tag{2}$$

Since we also have

$$\tilde{N}^{\lambda}{}_{\mu\nu} = N^{\lambda}{}_{\mu\nu} + \tfrac{2}{3}\delta^{\lambda}{}_{\mu}\Gamma_{\nu}, \tag{3}$$

then

$$\tilde{N}_{\mu\lambda\nu} = N_{\mu\lambda\nu} + \tfrac{2}{3}a_{\lambda\mu}\Gamma_{\nu}, \tag{4}$$

where

$$N_{\mu\lambda\nu} = a_{\lambda\sigma}N^{\sigma}{}_{\mu\nu}, \quad \tilde{N}_{\mu\lambda\nu} = a_{\lambda\sigma}\tilde{N}^{\sigma}{}_{\mu\nu}. \tag{5}$$

Hence

$$\tilde{N}_{[\mu\lambda]\nu} = N_{[\mu\lambda]\nu}. \tag{6}$$

Therefore, we can write the relation (2) in the form

$$\tilde{H}_{\lambda\mu\nu} = H_{\lambda\mu\nu} + \tfrac{2}{3}\Gamma_{\lambda}a_{\mu\nu} + A_{\lambda\mu\nu} \tag{7}$$

where the tensor $A_{\lambda\mu\nu}$ is skewsymmetric in its last index pair:

$$A_{\lambda\mu\nu} = -A_{\lambda\nu\mu}.$$

Consider now equations (2.21) and (2.22). We have

$$\sigma_{\mu}{}^{\dot{A}C}\sigma^{\nu}{}_{\dot{A}B}(N^{\mu}{}_{\nu\lambda} - H^{\mu}{}_{\lambda\nu}) = \sigma^{\mu\dot{A}C}\sigma^{\nu}{}_{\dot{A}B}(N_{\nu\mu\lambda} - H_{\lambda\mu\nu}), \tag{8}$$

but only the part which is skewsymmetric in μ and ν contributes to this expression.

To show this, observe that the first of the equations (2.4) can be written in the form

$$N_{\mu\nu\lambda} + N_{\nu\mu\lambda} = -W_{\lambda\mu\nu}. \tag{9}$$

Permuting the indices λ, μ, ν cyclically, adding the first two of the equations so obtained and subtracting the third, we find that

$$N_{(\mu|\nu|\lambda)} - N_{[\mu|\lambda|\nu]} - N_{[\lambda|\mu|\nu]} = -\tfrac{1}{2}(W_{\lambda\mu\nu} + W_{\mu\nu\lambda} - W_{\nu\lambda\mu}). \tag{10}$$

Hence

$$N_{\mu\nu\lambda} - H_{\lambda\nu\mu} = -\tfrac{1}{2}(W_{\lambda\mu\nu} + W_{\mu\nu\lambda} - W_{\nu\mu\lambda})$$
$$+ S_{\mu\lambda\nu} + S_{\lambda\mu\nu} - S_{\lambda\nu\mu} - H_{\lambda\nu\mu}, \tag{11}$$

where

$$S_{\mu\lambda\nu} = N_{[\mu|\lambda|\nu]} = a_{\lambda\sigma}N^{\sigma}_{[\mu\nu]} = a_{\lambda\sigma}\Gamma^{\sigma}_{[\mu\nu]} = -S_{\nu\lambda\mu}. \tag{12}$$

Symmetrising equation (11) over μ and ν,

$$N_{(\mu\nu)\lambda} - H_{\lambda(\mu\nu)} = \tfrac{1}{2}W_{\mu\nu\lambda} - H_{\lambda(\mu\nu)} = 0, \tag{13}$$

proving the result. In exactly the same way

$$\tilde{N}_{(\mu\nu)\lambda} - \tilde{H}_{\lambda(\mu\nu)} = 0. \tag{14}$$

Let us define also the tensors

$$T_{\lambda\mu\nu} = H_{\lambda\mu\nu} - S_{\nu\mu\lambda} \quad \text{and} \quad \tilde{T}_{\lambda\mu\nu} = \tilde{H}_{\lambda\mu\nu} - \tilde{S}_{\nu\mu\lambda}. \tag{15}$$

Elementary substitution now shows that

$$N_{\mu\nu\lambda} - H_{\lambda\nu\mu} = \tfrac{1}{2}[T_{[\lambda\mu\nu]} - T_{[\lambda\nu\mu]} + 2(T_{\mu\lambda\nu} - T_{\nu\lambda\mu})], \tag{16}$$

and

$$\tilde{N}_{\mu\nu\lambda} - \tilde{H}_{\lambda\nu\mu} = \tfrac{1}{2}[\tilde{T}_{[\lambda\mu\nu]} - \tilde{T}_{[\lambda\nu\mu]} + 2(\tilde{T}_{\mu\lambda\nu} - \tilde{T}_{\nu\lambda\mu})]. \tag{17}$$

It can be readily seen that equations (13) and (14) follow immediately from these two expressions.

We are now ready to discuss the minimum coupling hypothesis and the Dirac equations of electrodynamics in the generalised field theory.

4. A comment on the minimum coupling hypothesis

Spinor algebra and analysis must now be applied in a physical context. I shall show in the concluding sections of this chapter that our formalism leads to a particularly firm association between a Dirac electrodynamics and GFT. Let us start however with a brief comment on the standard method of writing the Dirac equations in a curved space–time.

The Dirac equations were, of course, originally designed as a Lorentz invariant set of first-order partial differential equations arising from the second-order Klein–Gordon equation describing massive particles in a reverse process to that by which the wave equation is obtained in Maxwell's theory. Since there is an almost complete lack of empirical evidence on the interaction between the gravitational field and microphysics, extension of the Dirac system to a curved space–time must be based on an *a priori* assumption. The most common is the so called hypothesis of minimum coupling whereby partial derivatives are replaced by covariant derivatives without changing any other terms in the equations. The hypothesis is a clear application of the principle of simplicity but its effect are virtually untestable.

The reason is primarily the local flatness of a Riemannian space–

time. It would be easy to envisage addition to the invariant equations of a term such as

$$R_{\mu\nu\lambda\kappa}\sigma^{\mu A\dot{C}}\sigma^{\nu}{}_{\dot{C}B}\sigma^{\lambda B\dot{D}}\sigma^{\kappa}{}_{\dot{D}E}\psi^{E} \tag{1}$$

which would automatically vanish in a flat space–time.

Moreover, in the classical theory additional assumptions foreign to macrophysics are involved in the transition to a case in which there is an interaction with an exterior electromagnetic field φ_{λ}:

$$\partial_{\lambda} \to \partial_{\lambda} + i\varphi_{\lambda}. \tag{2}$$

If one were to drop the assumption that γ_{AB} is a constant and write

$$\gamma_{AB} = \begin{bmatrix} 0 & 1 \\ -1 & 0 \end{bmatrix} \gamma e^{i\theta}, \tag{3}$$

then equations (2.18) can be readily seen to be invariant under the substitution

$$\begin{Bmatrix} A \\ \lambda B \end{Bmatrix} \to \begin{Bmatrix} A \\ \lambda B \end{Bmatrix} + i\varphi_{\lambda}\delta^{A}{}_{B}, \tag{4}$$

where φ_{λ} is an arbitrary real vector.

In a general relativistic theory, the gradient

$$\theta_{\lambda} = \partial_{\lambda}\theta \tag{5}$$

then corresponds to a gauge transformation. If the trace $\begin{Bmatrix} A \\ \lambda A \end{Bmatrix}$ is not pure imaginary

$$\begin{Bmatrix} A \\ \lambda A \end{Bmatrix} + \begin{Bmatrix} \dot{A} \\ \lambda \dot{A} \end{Bmatrix} = 2\partial_{\lambda}\ln\gamma, \tag{6}$$

and there is an obvious temptation to identify φ_{λ} with an electromagnetic vector potential. But the latter is arbitrary and therefore the procedure can hardly be consistent with the concept of geometrisation of an electromagnetic field.

The nongeometrical nature of φ_{λ} can be avoided in the generalised field theory by an identification

$$\varphi_{\lambda} = \Gamma_{\lambda}, \tag{7}$$

but this is no less arbitrary and we shall see presently that it is also unnecessary.

Before proceeding however to consider the Dirac equations in the curved space of GFT, let us digress once more to see what might be the meaning of a flat-space approximation in our extension of general relativity.

I am indebted for the formal introduction of spinors into the nonsymmetric theory to my student, B. T. McInnes.

5. Flat space–time in GFT

Let us suppose that the background space–time of GFT is a flat Minkowski space:

$$a_{\mu\nu} = \eta_{\mu\nu} = \text{diag}(+1, -1, -1, -1). \tag{1}$$

Then, globally

$$\tilde{\Gamma}^{\lambda}_{(\mu\nu)} = 0, \tag{2}$$

and, by the invariance of $\eta_{\mu\nu}$, the group of allowable coordinate transformations collapses into the Poincaré group. The field equations become (writing as before $g_{(\mu\nu)} = h_{\mu\nu}$, $g_{[\mu\nu]} = k_{\mu\nu}$)

$$h_{\mu\nu,\lambda} - \tilde{\Gamma}^{\sigma}_{(\mu\lambda)} k_{\sigma\nu} - \tilde{\Gamma}^{\sigma}_{(\lambda\nu)} k_{\mu\sigma} = 0$$

$$= k_{\mu\nu,\lambda} - \tilde{\Gamma}^{\sigma}_{[\mu\lambda]} h_{\sigma\nu} - \tilde{\Gamma}^{\sigma}_{[\lambda\nu]} h_{\mu\sigma}; \tag{3}$$

$$g = \text{constant}; \tag{4}$$

$$\tilde{\Gamma}^{\rho}_{[\mu\sigma]} \tilde{\Gamma}^{\sigma}_{[\rho\nu]} = 0; \quad R_{[\mu\nu]} = -\tilde{\Gamma}^{\sigma}_{[\mu\nu],\sigma} \tag{5}$$

The first of the equations in (3) implies twenty conditions

$$h_{[\mu\nu,\lambda]} = 0, \tag{6}$$

from which a straightforward calculation (e.g. considering h_{11}, h_{22} and h_{12}), shows that, if the g field is regular at infinity,

$$h_{\mu\nu} = \eta_{\mu\nu}. \tag{7}$$

Thus, a flat background necessarily (and as it should) implies the absence of a gravitational field. The remaining field equations give

$$\tilde{\Gamma}^{\sigma}_{[\mu\nu]} = \tfrac{1}{3}\gamma_{\nu}\delta^{\sigma}_{\mu} - \tfrac{1}{3}\gamma_{\mu}\delta^{\sigma}_{\nu} + \gamma^{\sigma}_{\mu\nu}, \tag{8}$$

$$\gamma^{\sigma}_{\mu\nu,\sigma} = 0, \tag{9}$$

and

$$\gamma^{\sigma}_{\mu\rho}\gamma^{\rho}_{\sigma\nu} + \tfrac{1}{3}\gamma_{\mu}\gamma_{\nu} = 0, \tag{10}$$

where

$$\gamma^{\sigma}_{\mu\nu} = -\gamma^{\sigma}_{\nu\mu} \quad \text{and} \quad \gamma_{\mu} = \gamma^{\sigma}_{\mu\sigma}.$$

Because of the second of the equations in (3), the tensor $\gamma^{\sigma}_{\mu\nu}$ is a linear functional of γ_{μ} and of $k_{\mu\nu,\lambda}$, while the first equation collapses into algebraic conditions

$$\tilde{\Gamma}^{\sigma}_{[\mu\lambda]} k_{\sigma\nu} = \tilde{\Gamma}^{\sigma}_{[\lambda\nu]} k_{\sigma\mu}. \tag{11}$$

Since by equations (4) and (7) the $k_{\mu\nu}$ field has only five independent components (up to an arbitrary g) and new, independent algebraic equations can be set up by differentiating equations (10) and (11), a simple count shows that in general

$$k_{\mu\nu} = 0 = \tilde{\Gamma}^{\lambda}{}_{\mu\nu}. \tag{12}$$

The fact, however, that

$$g_{\mu\nu} = \eta_{\mu\nu}$$

does not mean that we have got rid of the physical content of the theory. Since we are in a Minkowski space–time, the connection

$$\Gamma^{\sigma}{}_{(\mu\nu)}$$

is clearly spurious but now the tensor

$$\Gamma^{\sigma}{}_{[\mu\nu]} = \tfrac{1}{3}(\delta^{\sigma}{}_{\nu}\Gamma_{\mu} - \delta^{\sigma}{}_{\mu}\Gamma_{\nu}) \tag{13}$$

satisfies the equation

$$R_{[\mu\nu]} = -\Gamma^{\sigma}{}_{[\mu\nu],\sigma} = \tfrac{1}{3}(\Gamma_{\nu,\mu} - \Gamma_{\mu,\nu}) \tag{14}$$

and gives the intensity set of the electromagnetic field equations if $\tfrac{1}{3}\Gamma_{\lambda}$ is identified with the electromagnetic vector potential, and $R_{[\mu\nu]}$ with the field tensor. It may be recalled (equations (2.3.17)) that previously the vector potential was identified as $-\tfrac{2}{3}\Gamma_{\lambda}$, but since the identification is made only up to a constant factor it seems untimely to speculate whether this difference could be ascribed to the absence of a gravitational interaction.

Of more importance is the fact that GFT has now been shown to be reducible in the absence of gravitation to a flat-space electromagnetic theory. I shall discuss this aspect of the theory again in the final chapter. At the moment, let us observe that this is no more than could be expected as far as microphysical Dirac theory is concerned.

6.   **Dirac equations and the metric hypothesis**
        In view of the above, it follows that the curved space analogues of the Dirac equations should be written in such a way that the introduction of an exterior electromagnetic interaction should be contained in their structure. I shall now show that this leads not only to a unique form of the Dirac equations but also to the metric hypothesis

$$\left\{ \begin{matrix} \lambda \\ \mu\nu \end{matrix} \right\}_{a} = \tilde{\Gamma}^{\lambda}{}_{(\mu\nu)}$$

on which GFT is founded.

Indeed, let us consider the equations for a (spin $\tfrac{1}{2}$) particle of rest mass m in the form

$$\sqrt{2}\sigma^{\mu\dot{A}B}\nabla_{\mu}\psi_{B} + m\dot{\varphi}^{A} = 0, \tag{1}$$

$$\sqrt{2}\sigma^{\mu\dot{A}B}\nabla_\mu\varphi_{\dot{A}} + m\psi^B = 0, \tag{2}$$

equation (2) being the conjugate of (1). There is little choice in GFT apart from the above generalisation of the flat-space Dirac equations. The possible alternative with $\tilde{\nabla}$ replacing ∇ does not give the correct electromagnetic or spin interaction terms. We may note that a 'minimum coupling' hypothesis is not involved in writing down equations (1) and (2) because of the initial separation of geometry and physics.

From the results of section 4 we have

$$N^\mu{}_{\nu\lambda} - H^\mu{}_{\lambda\nu} = \tilde{N}^\mu{}_{\nu\lambda} - \tilde{H}^\mu{}_{\lambda\nu} - A^\mu{}_{\lambda\nu} \tag{3}$$

and equation (1) (which is all that we need to consider) becomes

$$\sqrt{2}\sigma^{\lambda\dot{A}B}(D_\lambda\psi_B - \tfrac{1}{2}\sigma^{\mu\dot{D}C}\sigma^\nu{}_{\dot{D}B}(\tilde{N}_{[\nu\mu]\lambda} - \tilde{H}_{\lambda[\mu\nu]})\psi_C$$
$$+ \tfrac{1}{2}\sigma^{\mu\dot{D}C}\sigma^\nu{}_{\dot{D}B}A_{\lambda\mu\nu}\psi_C + \tfrac{1}{2}\Gamma^{\dot{D}}{}_{\lambda\dot{D}}\psi_B) + m\varphi^{\dot{A}} = 0, \tag{4}$$

because of the solution (2.21).

A natural way of avoiding ambiguity in whether tilded or untilded operators should be employed, is to assume that

$$\nabla_\lambda\gamma_{AB} = u\Gamma_\lambda\gamma_{AB}, \tag{5}$$

where u is a numerical, real or complex, multiplier. Since

$$\tilde{\Gamma}_\lambda \equiv 0,$$

it then follows that

$$D_\lambda\gamma_{AB} = 0 = \tilde{\nabla}_\lambda\gamma_{AB}, \tag{6}$$

and therefore, with a constant fundamental spinor γ_{AB}, that

$$\left\{ \begin{matrix} A \\ \lambda A \end{matrix} \right\} = \tilde{\Gamma}^A{}_{\lambda A} = 0. \tag{7}$$

The second of the equations in (2.9):

$$\nabla_\lambda(\gamma_{\dot{A}\dot{B}}\gamma_{CD}) = (\nabla_\lambda\gamma_{\dot{A}\dot{B}})\gamma_{CD} + \gamma_{\dot{A}\dot{B}}\nabla_\lambda\gamma_{CD} = 0,$$

will automatically be satisfied if

$$\nabla_\lambda\gamma_{\dot{A}\dot{B}} = -u\Gamma_\lambda\gamma_{\dot{A}\dot{B}}. \tag{8}$$

We are thus led to the conclusion that ∇_λ must be an anti-Hermitian operator and

$$u = iw, \quad w \text{ real.} \tag{9}$$

In this case, the equation

$$\nabla_\lambda\gamma_{AB} = -\Gamma^C{}_{\lambda A}\gamma_{CB} - \Gamma^C{}_{\lambda B}\gamma_{AC} = iw\Gamma_\lambda, \tag{10}$$

implies that

$$\Gamma^A{}_{\lambda A} = -iw\Gamma_\lambda, \tag{11}$$

so that, if also

$$\tilde{N}_{[\nu\mu]\lambda} - \tilde{H}_{\lambda[\mu\nu]} = 0, \tag{12}$$

the Dirac equation (4) becomes

$$\sqrt{2}\,\sigma^{\lambda\dot{A}B}(D_\lambda\psi_B + \tfrac{1}{2}\sigma^{\mu\dot{D}C}\sigma^\nu{}_{\dot{D}B}A_{\lambda\mu\nu}\psi_C + \tfrac{1}{2}\mathrm{i}w\Gamma_\lambda\psi_B) + m\varphi^{\dot{A}} = 0. \tag{13}$$

The term

$$\tilde{N}_{[\nu\mu]\lambda} - \tilde{H}_{\lambda[\mu\nu]}$$

represents in GFT an electromagnetic coupling which is already expressed by Γ_λ. Hence there is no less of generality in assuming that it should vanish. The lack of any further interaction with the electromagnetic field apart from the appearance of the vector potential is then in complete agreement with the empirically well-tested quantum electrodynamics. Also the term containing $A_{\lambda\mu\nu}$ expresses now the spin properties of matter or rather of a particle-electron. This term is a correction to the standard Dirac equation. Because of empirical evidence, it is presumably very small and there may be little harm in supposing that

$$A_{\lambda\mu\nu} = 0, \tag{14}$$

although the theory does not demand this. If however, we assume that $A_{\lambda\mu\nu}$ should vanish, the relation (12) can likewise be strengthened to

$$\tilde{N}_{\nu\mu\lambda} - \tilde{H}_{\lambda\mu\nu} = 0. \tag{15}$$

Let us now consider equation (3.17). Defining the density

$$\mathfrak{T}^\alpha = \varepsilon^{\alpha\lambda\mu\nu}\tilde{T}_{\lambda\mu\nu}, \tag{16}$$

we get from (15)

$$6\mathfrak{T}^\alpha + 2(-\mathfrak{T}^\alpha - \mathfrak{T}^\alpha) = 2\mathfrak{T}^\alpha = 0,$$

whence

$$\tilde{T}_{\mu\lambda\nu} - \tilde{T}_{\nu\lambda\mu} = 0, \tag{17}$$

or

$$\tilde{H}_{\lambda\mu\nu} - \tilde{H}_{\nu\mu\lambda} = 2\tilde{S}_{\nu\mu\lambda}. \tag{18}$$

It now follows from the definition (2.23) that

$$\tfrac{1}{2}(\tilde{H}_{\lambda\mu\nu} + \tilde{H}_{\nu\mu\lambda}) = a_{\mu\sigma}\left(\tilde{\Gamma}^\sigma{}_{(\nu\lambda)} - \left\{\begin{matrix}\sigma\\\nu\lambda\end{matrix}\right\}_a\right). \tag{19}$$

The metric hypothesis is thus equivalent to the plausible condition

$$\tilde{H}^\mu{}_{\lambda\nu} = \tilde{H}^\mu{}_{[\lambda\nu]}. \tag{20}$$

We should stress also that the assumption of skewsymmetry of the H

tensors enables them to be uniquely determined. This requirement is reasonable because of their crucial role in the spinor analysis of the nonsymmetric theory (equation (2.5)).

If the background space–time is Minkowskian, the Dirac equations (1) and (2) iterate (under the assumption (14)) to the standard, Schrödinger–Klein–Gordon equation of the second order. This again reinforces our choice of them as the correct description of the spin $\frac{1}{2}$ particles.

The scalar w is then identified as

$$w = \tfrac{3}{4}e, \tag{21}$$

where e is the (electronic) charge which gives rise to the external electromagnetic field.

This completes our account of the relationship between GFT and microphysics although other possibilities will be mentioned briefly in the last chapter of this monograph.

8 An outlook

1. Generalised field theory as a completion of relativity

The theory outlined in the preceding pages represents a natural extension of general relativity, just as the latter modified and widened the assumptions of special relativistic physics through replacing equivalence of uniformly moving observers by that of observers in arbitrary relative motion. This is the content of the principle of covariance. Nevertheless there is a significant difference between special relativity and GR. The ideas introduced in the former found an application throughout all branches of physics. On the other hand, as a result of the narrow imposition of the principle of equivalence, general relativity became little more than a theory, however good, of the gravitational field. It is important to note that it has been empirically confirmed only in this role and then only as far as the solutions of the empty field equations are concerned.

From a physical point of view, the general relativistic field equations with sources, that is with a nonzero energy–momentum tensor $T_{\mu\nu}$, do little more than determine the gravitational field corresponding to a given energy distribution. They certainly do not provide us with a theory of any nongravitational field so that even if the resulting geometrical structure were observed (an unlikely event) we would be at a loss to know whether the equations have been correctly applied or not. The reason is that the requirement of covariance alone does not enable us to construct $T_{\mu\nu}$ whose form can only be guessed on prerelativistic grounds which have nothing to do with the theory itself. This inability of GR to handle anything other than gravity, and then in a virtually untestable way, was recognised by Einstein. It was the starting point of his unified field theory of which GFT is the outcome.

Einstein himself emphasised the nonrelativistic nature of any assumed form of $T_{\mu\nu}$ rather than the possible nonverifiability of the resulting theory, though no doubt this is what he had in mind. In particular, he was concerned about the failure of GR to describe adequately the electromagnetic field. Two theories must be mentioned in this context.

The first is the Einstein–Maxwell theory when $T_{\mu\nu}$ becomes the Maxwell energy–stress–momentum tensor. It leads to the standard Reissner–Nordström solution for a stationary, spherically symmetric charge and all the relativistic models of things such as charged, rotating black holes. I doubt if any observational evidence can sustain the claim to verify (or disprove) such objects. But the main fault of the Einstein–Maxwell theory is that it virtually presupposes Maxwell's electrodynamics and hardly advances our understanding of physics. Of course, Maxwell's theory and, perhaps even more so, quantum electrodynamics, are very well tested indeed but only locally and not in connection with a curved space–time. Presupposition of classical electromagnetism is also the characteristic of Rainich's already unified field theory (and to some extent of Weyl's unified field theory, a discussion of which is beyond the scope of this monograph). Rainich's theory is simply Einstein–Maxwell's with the energy–momentum tensor eliminated by algebraic means. It met with the destructive objection that it implied an unobserved relation between charge and mass (see e.g. Klotz & Lynch 1970).

This objection can be applied also to the Einstein–Maxwell theory in the sense that it demands a weak electromagnetic field to correspond necessarily to a weak gravitational field which alone in GR determines geometry of the space–time and hence the laws of measurement. I have referred already (chapter 7, section 5) to the fact that GFT reduces to special relativity (i.e. to classical electromagnetism) in the absence of gravitation or, rather, when the background space–time is flat. Such reducibility can often be expected of generalisations of existing theories; for example SR mechanics becomes Newtonian if the speed of light is allowed to tend to infinity. GR does not satisfy the reducibility requirement, since the flat-space limit of it is emphatically empty. On the other hand, it might be objected that our 'residual' electromagnetism contradicts the empirical facts of the Pound–Rebka experiment. I do not think so. The experiment is an attempt to verify the equivalence principle in a gravitational field and it is well known that SR to which generalised field theory reduces in a flat background is incapable of describing the effects of gravity correctly.

The most characteristic failures of the numerous proposals for a unified field theory can be summarised as two-fold. One is that almost invariably they set off from the standard description of gravitation and electromagnetism and merely sought a mnemonic device for combining them formally. The result was an untestable theory. The second, and somewhat more subtle fault, was that they sought an *a priori*, and impossible to justify, generalisation of geometry, forgetting that very basic mathematical, but physically meaningful, reasons exist (paracompactness) which demand a Riemannian space–time as the vehicle for physical events.

I now claim that GFT is free from these deficiencies. In a sense, the reservation about geometry-extending theories can be made also about the recent, quantum oriented unifications despite their apparent empirical success and therefore, well-deserved recognition. We can describe them in short as model dependent, whether that model is a five-dimensional geometry or a particular gauge-invariance group. Contrariwise, GFT is derived from a physically comprehensible hypothesis from which the model, that is the theory itself, is constructed on the basis of a general argument.

Einstein saw that this hypothesis had to replace the general relativistic equivalence principle and that the structure of the 'total' macroscopic field should somehow contain 'something like' Maxwell's electromagnetism. That it should be electromagnetic theory followed from the conviction that like gravitation, electromagnetic field ought to determine geometry on equal footing and not merely because it happened to be energy carrying and thus a source of space–time curvature. GFT turned out to contain not only Maxwell's theory, albeit in a modified, nonlinear form underlining the 'something like' aspect, but also Dirac's electrodynamics. This however, did not become apparent until the foundations of the theory had been completely revised.

In the initial stages, in which Einstein left his unified field theory, it was perhaps impossible to avoid what is now known to have been a false interpretation to which most of the troubles of the theory (I shall summarise these in the next section) were due. In particular, Einstein's field equations were hypothetical, guessed rather than derived. It can only be regarded as a tribute to his genius that these equations, the weak field equations, should be the most general (within reason) set consistent with the principle of charge conjugation on Hermitian symmetry. What Einstein could not recognise when the theory was being set up was that the field equations by themselves were insufficient to determine the geometry of the space–time, nor that the latter could still remain Riemannian. Neither could it be seen that their dependent functions (the components $g_{\mu\nu}$ of the fundamental tensor) had to stand for physical fields which in interaction should not be directly related to geometrical concepts as in general relativity.

The crucial step was the metric hypothesis. Indeed, far from being an *a priori* assumption as its somewhat unfortunate name might suggest, it was a consequence of the requirement of consistency with the original principle of Hermitian symmetry which links the mathematical structure of GFT with physics. The principle may well have to be changed in future but only if the theory described here should prove empirically untenable, or perhaps if 'new' macrofields are discovered. As it stands, it is the basis of a completion of general relativity, that is of an extension of its concepts to nongravitational fields. I have said several times throughout this monograph that the energy–momentum tensor should be eliminated from the field equations describing the interaction of fundamental fields, and that it must be calculated after the field equations are solved and their solution interpreted physically. But what fields are 'fundamental' or, rather, what is the justification for regarding electromagnetism and gravitation as the only fundamental macrophysical fields? They are certainly the only ones known and I have already mentioned misgivings of unqualified inclusion into the unified structure of basically quantum mechanical concepts. Equally certain is the conceptual necessity of extending GR to contain, and to describe structurally, the electromagnetic field. Again, it is in this sense that GFT is to be regarded as a completion of general relativity. If therefore the fields related to the weak and strong interactions are left aside; not because they are unimportant, nor on account of their range, but because their description necessarily involves ideas which can not be at present incorporated meaningfully into a continuous picture of reality; this completion is limited to the two macrophysical fields we have been considering. This much is almost *a priori* written into GFT by regarding as fundamental a second rank tensor $g_{\mu\nu}$ with only sixteen components. Nothing more is required if one starts from a general relativistic point of view.

2. Objections to the nonsymmetric theory

The aim of this monograph is to present the generalised field theory as a logically consistent account of macrophysics whose immediate consequences conform to known physical facts or expectations. I shall discuss in the next section the vital question of the sense in which the theory leads to predictions susceptible to empirical verification. GFT, like any theory, becomes physical not as a result of its structure but on account of the interpretation we care to put on the quantities with which it operates.

As an example, we have met (chapter 4) the conditions which the electromagnetic field tensors must satisfy if the equations of motion of a charged test particle derived by the EIH method are to contain a Lorentz force. It is one of the 'expectations' of a nonlinear theory (with Bianchi identities) that these equations should be obtainable from the field equations without additional, physical hypotheses. This does not mean, of course, that mathematical assumptions (e.g. a method of approximation with only peripheral physical content) are not made in their derivation.

Another expectation which our intuitive, though strictly macro-physical, understanding of electromagnetism implies is that the theory should admit existence of spherically symmetric charges. This is fulfilled by GFT in a surprising manner, although the requirement itself is but a corollary of wanting 'something like' Maxwell's theory to emerge. The field of a static, structureless charge is exactly, that is without a 'testable' correction, a Coulomb field.

Both the solution of the problem of motion and the existence of spherically symmetric 'point'-charges implicitly answer two major objections to the nonsymmetric unified field theory. Infeld in 1950 (considering strong field equations) and Callaway three years later (for WFE which are the basis of current theory) showed that Lorentz force appears in a lower order of approximation than a gravitational force. Their results, however, were obtained starting from the definition of $g_{[\mu\nu]}$ as the electromagnetic field. We have seen that this is inconsistent not only with the problem of motion they tackled but also with the interpretation of the static solution. The correct interpretation pro-duced not only a Lorentz force but in addition disposed of an objection due to Tiwari & Pant (1970) who could not find a solution correspond-ing to a stationary, spherically symmetric charge.

A. Papapetrou has drawn my attention to a third objection which appears to have been largely ignored in the literature although Einstein himself attempted to answer it immediately when it was raised. It concerns the interpretation of similarity solutions $g(kx)$ for a con-tinuously varying parameter k (if $g(x)$ is a solution of the field equations then $g(kx)$ will be a solution too). C. P. Johnson, author of the objection, argued that for a system of two charged particles and one uncharged particle in equilibrium according to the $g(x)$ solution, $g(kx)$

results in relative motion, charges varying as k^2 and masses as k. The conclusion was based on Einstein's definition of charge density as

$$\rho = \tfrac{1}{2}\varepsilon^{0ijk} g_{[ij],k}. \tag{1}$$

In other words, it looked as if the nonsymmetric field theory necessarily violated the Coulomb law.

Einstein's reply consisted of two points. The first laid down a curious requirement, apparently derived from the atomistic structure of matter, that an acceptable theory should exclude coexistence of similarity solutions. Then, with his customary insight of genius, he suggested (since he was not in a position to demonstrate it explicitly) that the transformation

$$k \to kx \tag{2}$$

corresponds to a transition to a different world.

That this indeed is the case follows immediately from the cosmological interpretation of chapter 6, of the metric

$$ds^2 = \gamma dt^2 - \gamma^{-1} w^2 dr^2 - wr^2 d\Omega^2, \tag{3}$$
$$\gamma = 1 - 2m/r, \quad w = r_0^2/r_0^2 + r^2).$$

The similarity solution corresponds to a new r_0 (radius of the universe) and m ('mass-radius' of the primeval but presumably recurring fire-ball) and there is no reason to suppose that particles in equilibrium in one world should remain in equilibrium in another. At least, one can not be sure of this without proof.

Cosmological interpretation of the metric disposes *prima facie* of Johnson's objection but it is expedient to look at it more closely to see what actually happens. And it is not necessary to have even a three-particle system if the equilibrium of two bodies (that is the balance between the Coulomb and Newton forces) can be disturbed by a similarity transformation.

We observe first that the Newtonian gravitational constant N, which is not dimensionless, remains invariant under a similarity transformation only if mass is assumed to transform in the same way as the space–time coordinates (length). This assumption is often concealed. For example, in GR with the mass density

$$\rho \propto T_{00} \tag{4}$$

the field equations ($T_{\mu\nu}$ being proportional to second derivatives of the metric tensor) give a variation as k^2 of ρ – equivalent to our assumption. A similar variation of the charge density is implied by Maxwell's theory. It is not surprising therefore that Einstein–Maxwell theory does not exhibit the Johnson paradox.

But neither does GFT if instead of postulating more or less arbitrary variation of physical quantities (and in particular of the charge and mass densities) induced by a similarity transformation, it is remembered that even statical balance of forces is determined by the equations of motion. They are the only observable characteristics of the situation, since, apart from a dimensionally irrelevant correction (the terms linear in r), the GFT equations of motion of a charged massive test particle are classical. Suppose now that the Newtonian and Coulombian forces balance for a $g(x)$ solution:

$$Nmm'/r^2 = ee'/r^2. \tag{5}$$

Then with

$$x = k\bar{x}, \quad m = k^\alpha \bar{m}, \quad N = k^{1-\alpha}\bar{N}, \quad e = k^\gamma \bar{e}, \tag{6}$$

(the bar representing similarity transformed quantities), the acceleration $\bar{a}$ which might arise in a $g(kx)$ solution (which we could have called $\bar{g}$) is given by

$$\bar{m}\bar{a}r^2/ee' = k^{\alpha-1}(k^{2(\gamma+1-\alpha)} - 1), \tag{7}$$

and vanishes if and only if

$$\gamma + 1 - \alpha = 0. \tag{8}$$

Now, in GFT, the vector

$$\Gamma_\mu = \Gamma^\sigma_{[\mu\sigma]} \tag{9}$$

is identified up to a numerical factor, with the electromagnetic vector potential. Hence the charge remains invariant:

$$\gamma = 0, \tag{10}$$

and therefore, as the necessary consequence of the equations of motion, acceleration will vanish also for the transformed solution when $\alpha = 1$ (and the Newtonian constant remains unchanged in the new universe).

3. The possibility of an empirical test of GFT

We have seen (chapter 4) that integration of the field equations in an EIH approximation implies existence of a Treder potential

$$\left(\frac{d^2}{dr^2} + \frac{2}{r}\frac{d}{dr}\right)^2 \varphi = 0, \tag{1}$$

except that φ is not the electrostatic potential of the field. On the contrary, the field of a stationary point-like charge is strictly

$$e\hat{r}/r^2 \tag{2}$$

the inverse square, Coulomb field. Moreover, close to such a source (that is if r_0 can be regarded as infinite), the space–time becomes Schwarzschild. The deviation represented by the solution (5.6.4) is exceedingly small and unlikely to be detectable in practice. The same can be said about the cylindrical correction of chapter 4, section 8. In other words, it is almost impossible to visualise an empirical verification of GFT to be carried out on a local scale on the basis of the primitive solutions hitherto discovered. It looks as if, locally, general relativity and electromagnetism part company; the gravitational and electromagnetic fields do not interact except perhaps in the regions of very high energy as indicated by the current, quantum based unified theories, or perhaps where very strong charge is present. The latter however, is unlikely to remain stationary in an experiment and high energy regions can hardly be regarded as geometrically local because of the implied curvature of the space–time. Any deviation due to the nonlinearity of electrodynamics predicted by GFT, is again of a nonlocal character or requires excessively strong fields to show up against the classical electrodynamics of Maxwell.

Local bifurcation of electromagnetism and gravitation is not unexpected. Theoretically speaking, the metric relation (5.6.4):

$$ds^2 = \gamma dt^2 - \frac{r_0^2 dr^2}{\gamma(r_0^2 - r^2)} - r^2 d\Omega^2,$$

$$\gamma = 1 + c \sqrt{(r_0^2/r^2 - 1)},$$

is the GFT counterpart of the Reissner–Nordström solution of the Einstein–Maxwell theory. Nevertheless, one can not expect even a valid theory to show large, or even large enough to be locally observable, effects contradicting locally established theories (GR and Maxwell). Hence the most likely test of GFT may come not from its consequences close to an observer, but from its global predictions.

It is for this reason that one can not overemphasise the importance and novelty of the cosmological implications of generalised field theory. The latter is the only theory of its kind which leads to a cosmological model without additional assumptions other than interpretative but this is hardly a drawback. An assumption is necessarily involved when a mathematical equation or solution is related to physical reality. What matters is that the model of chapter 6 was derived not by writing a cosmological principle into the hypotheses of the solution, but in an effort to make sense of a Coulomb field apparently cutting off at a finite distance from its source. Therefore the resulting model can be referred to (as it was) without reservations as a cosmology without a cosmological principle.

Its empirical importance as far as GFT and its possible verification are concerned consists in the fact that the model is unique. It appears more sensible than its likewise 'principle-free' counterpart, the de Sitter universe, in that it does not demand an empty space–time. Matter indeed is present as we have seen, and moreover, it has a well-determined distribution. This at any rate could perhaps be investigated observationally and provide a test of the theory. The results of such an investigation should be theory independent unlike similar observational evidence in GR models where it is difficult to decide whether observations confirm the theory or any assumption about the postulated form of an energy–momentum tensor.

It must be noted however, that the suggested test of GFT from its cosmological implications need not be 'crucial' in the sense that its failure to conform to empirical reality should necessarily invalidate the theory. The structure of the universe was predicted on the basis of only a static, spherically symmetric solution of the field equations and of the metric problem. It is of course, the only general solution known (apart from plane symmetry solutions not considered in this monograph, and some partial results for a cylindrical field). It is possible that other, perhaps less restricted, solutions may lead to different cosmological models. It would then be a question of deciding which initial symmetry conditions best fit the reality.

On the other hand this is unlikely not so much because such solutions may not be found, but because it is difficult to imagine global situations in which the spherically symmetric case would not be meaningful.

Finally, if we return for a moment to local, terrestrial or laboratory predictions, perhaps the only significant one is the explanation of why spherically symmetric, magnetic monopoles can not exist. It is simply that they violate the conditions of the unified theory.

4.   Unsolved problems

The generalised field theory developed in this monograph is a completion of general relativity. Indeed, it is relativity freed from an energy–momentum tensor at the expense of a thorough revision of the basic concepts on which it is founded. Logical necessity of such a revision was unforeseen by Einstein. Therefore, it is both surprising and gratifying that the theory which he initiated should turn out to be the most general consistent with the hypothesis and Einstein's interpretation of the principle of Hermitian symmetry. Einstein was in a sense right after all! Other theories are possible but only if his principle is rejected or modified. They should be considered if GFT is shown by empirical evidence to give a wrong account of physical reality.

I have tried to carry GFT to the point where it might be possible for the evidence to decide its validity as a theory of physics. The result it that several major questions had to be left in abeyance. I shall attempt to list them now as they appear to me.

The most obvious problem is the scarcity of exact solutions on which the interpretation of the theory depends. In particular, axially symmetric solutions may conceivably lead to alternative cosmologies and of course, there is the need to compare them with the Kerr metric. Considerable effort has been devoted to the cylindrical case as an introduction to the more general problem. In fact it was these investigations that led to the interpretation of the electromagnetic field enabling the problem of motion, the main stumbling block of Einstein's original theory, to be solved.

The cosmological model discussed in this monograph shows the evolutionary features implied by observational evidence but it came from a static solution of the field equations. Even if spherical symmetry is taken for granted, it is possible that GFT admits time dependent solutions which could lead to alternative universes. In other words, it is an urgent problem to discover whether Birkhoff's theorem still holds in the generalised theory. I suspect that it does, in which case our model would be unique. If in addition its oscillatory nature is confirmed, it seems less to the point to worry about the early state of the universe. In general relativistic cosmologies these discussions involve also, and at best, a considerable degree of speculation which can only be verified by extensive extrapolation of concepts.

As far as electrodynamics is concerned, there are two major outstanding questions. The first is the precise meaning of the tensor

$$w_{\mu\nu} = g^{\alpha\beta}_{00} g_{[\mu\nu];\alpha\beta}, \tag{1}$$

which, along with $R_{[\mu\nu]}$ selected herein as the field tensor, satisfies the condition that it should give a Lorentz force on a charged test particle. For a spherically symmetric solution of the weak field equations, it retains one component viz.

$$w_{23} \propto (1/r^2)(1 - 2m/r) \tag{2}$$

Hence, if $w_{\mu\nu}$ were the intensity field, the Coulomb field would have a correction term. This would mean that the constant $2m$ should have an absolutely small value instead of just having

$$2m \ll r_0, \tag{3}$$

(because of the accuracy to which the inverse square law is empirically established). I chose $R_{[\mu\nu]}$ instead of $w_{\mu\nu}$ not so much because of the Coulomb law (although this of course, gave the immediate motivation for my cosmological argument) but on account of the relation

$$R_{[\mu\nu]}(\tilde{\Gamma}) \sim (\Gamma_{\nu,\mu} - \Gamma_{\mu,\nu}) \tag{4}$$

according to which $R_{[\mu\nu]}$ is exactly a Maxwell tensor.

I have already mentioned the second unsolved problem of electrodynamics of GFT and that is its connection with the theories of the Born–Infeld type. Investigation of this connection gives the only chance of a local or laboratory test of the theory, however slim that chance may be. It is also quite likely that the two problems will be resolved simultaneously.

Resolution of the problem of motion by a reinterpretation of the electromagnetic field is one of the triumphs of GFT. It re-established it as a viable theory even if the theory itself does not fulfil all of Einstein's expectations. I have expressed doubts on whether it can, or indeed whether it should, be required to describe the structure of matter (i.e. elementary particles) as Einstein thought. At any rate, this is clearly unlikely without the almost impossible task of revising the foundations of quantum mechanics. On the other hand, the relation of GFT to microphysics has so far been only cursorily investigated (chapter 7).

The problem of motion has been tackled only by approximate methods of the EIH technique. It would be interesting to know whether, for example, fast approximations can be adapted to the concepts of GFT. Similarly, the exact form of the equations of motion of a test particle is unknown. Russell and I have suggested once something like

$$\frac{\mathrm{d}^2 x^\alpha}{\mathrm{d}s^2} + \tilde{\Gamma}^\alpha{}_{(\mu\nu)} \frac{\mathrm{d}x^\mu}{\mathrm{d}s} \frac{\mathrm{d}x^\nu}{\mathrm{d}s} = \frac{1}{2} \frac{e}{m} \varepsilon^{\alpha\lambda\mu\nu} g_{(\lambda\kappa)} R_{[\mu\nu]} \frac{\mathrm{d}x^\kappa}{\mathrm{d}s}, \tag{5}$$

but of course, this form is purely hypothetical. For example, it could be $a_{\lambda\kappa}$ instead of $g_{(\lambda\kappa)}$, or $w_{\mu\nu}$ instead of $R_{[\mu\nu]}$ on the right hand side, too! Finally, the problem of gravitational waves in GFT requires consideration especially in view of the current empirical interest.

These then are the outstanding problems of the generalised field theory. If it survives the immediate tests, other problems will no doubt arise. I hope to have shown that GFT deserves serious consideration as a viable theory of macrophysics. As such it should stand as another monument to the insight of Albert Einstein. And whatever its fate as a theory of physics, GFT is the most consistent extension of relativistic thinking to embrace nongravitational fields. It is the completion of general relativity on the basis of the principle of Hermitian symmetry or of charge conjugation invariance.

Bibliography

Since this monograph is essentially an account of the work carried out by myself and my collaborators, very few references other than to our papers, which are listed below, seem to be appropriate.

Books

Carton, E. & Einstein, A. (1979) *Letters on Absolute Parallelism 1929–1932*, Princeton.

Eddington, A. S. (1924) *Mathematical Theory of Relativity*, Cambridge.

Einstein, A. (1951, 1954) *Meaning of Relativity*, London.

Hawking, S. W. & Israel, W. (eds.) (1979) *General Relativity, an Einstein Centenary Survey*, Cambridge.

Hlavaty, V. (1957) *Geometry of Einstein's Unified Field Theory*, Groningen.

Kobayashi, S. & Nomizu, K. (1963) *Foundations of Differential Geometry*, London.

Misner, C. W., Thorne, K. S. & Wheeler, J. A. (1973) *Gravitation*, San Francisco. (A comprehensive account of the growth of general relativity of which generalised field theory is the development.)

Spivak, M. (1970) *A Comprehensive Introduction to Differential Geometry*, Waltham, Mass.

Articles

Bade, W. L. & Jehle, H. (1953) *Rev. Mod. Phys.* **25**, 714.

Born, M. (1937) *Ann. Inst. H. Poincaré*, **7**, 155.

Callaway, I. (1953) *Phys. Rev.* **92**, 1567.

Einstein, A. (1945) *Ann. Math.* **46**, 578.

Einstein, A. & Infeld, L. (1949) *Can. J. Math.* **1**, 209.

Einstein, A. & Kaufman, B. (1955) *Ann. Math.* **62**, 128.

Einstein, A. & Straus, E. G. (1946) *Ann. Math.* **47**, 731.

Infeld, L. (1950) *Acta phys. pol.* **10**, 289. 1950.

Infeld, L. & Plebanski, I. (1954) *Proc. R. Soc. Lond.* **224**, 222.

Johnson, C. P. (1953) *Phys. Rev.* **89**, 329.

Mie, G. (1912, 1913) *Ann. Phys.* **37**, **39**, **40**.

Papapetrou, A. (1948) *Proc. R. Irish Acad.* **52**, 69.
Schrödinger, E. (1947) *Proc. R. Irish Acad.* **51**, 163.
Tiwari, R. & Pant, D. N. (1970) *Phys. Lett.* **33A**, 505.
Tonnelat, M. -A. (1955) *J. Phys. Radium* **16**, 21.
Vanstone, I. R. (1962) *Can. J. Math.* **14**, 568.
Wyman, M. (1950) *Can. J. Math.* **2**, 427.

Articles by A. H. Klotz and his collaborators relevant to the development of GFT
These articles are listed in chronological order, and the titles have been abbreviated to convey only their relevance to parts of the monographs.
(1969) Philosophy of Unified Field Theory, *Stud. Genet.* **22**, 1189.
(1970) On Hermitian Symmetry, *Acta phys. pol.* **B1**, 261.
(with I. Lynch; 1970) Is Rainich's Theory Tenable? *Nuovo Cim. Lett.* **4**, 248.
(with G. K. Russell; 1972) New Interpretation of UFT, *Acta phys. pol.* **B3**, 407.
(with G. K. Russell: 1972) Equations of Motion in UFT, *Acta phys. pol.* **B3**, 649.
(with G. K. Russell; 1973) Structure of Nonsymmetric Theory, *Acta phys. pol.* **B4**, 579.
(with G. K. Russell; 1973) Motion in a Cylindrical Field, *Acta phys. pol.* **B4**, 589.
(with L. J. Gregory; 1977*a*) Solution of Affine Equation, *Acta phys. pol.* **B8**, 595.
(with L. J. Gregory; 1977*b*) On the Electromagnetic Tensor, *Acta phys. pol.* **B8**, 601.
(1978) On the Nonsymmetric Theory, *Acta phys. pol.* **B9**, 573.
(1978) On Conservation Laws, *Acta phys. pol.* **B9**, 589.
(1978) Cosmological Consequences of Nonsymmetric Theory, *Acta phys. pol.* **B9**, 595.
(1979) New Cosmology, *Acta phys. pol.* **B10**, 295.
(1979) Geometry at Infinity, *Acta phys. pol.* **B10**, 307.
(1979) Cosmology without a Cosmological Principle, *Nuovo Cim. Lett.* **25**, 190.
(1979) Energy Tensor in the Nonsymmetric Theory, *Acta. phys. pol.* **B10**, 469.
(with B. T. McInnes; 1980) Meaning of Metric Hypothesis, *Acta phys. pol.* **B11**, 345.
(1980) Flat Space and Nonsymmetric Theory, *Acta phys. pol.* **B11**, 779.

(1980) Unified Field, Metric and Local Invariance Group, *Acta phys. pol.* **B11**, 501.

(1981) A Note on the Geometry of the New Cosmology, *Acta phys. pol.* **B12**, 11.

In *Proceedings of the 9th Conference on General Relativity*, Jena, July, 1980.

Index

affine bundle 44, 47, 48
affine connection 10, 19, 26, 28, 39

Bianchi identities 18, 32, 35, 36, 37, 53, 55, 56, 61, 139
bifurcation of geometry and physics 13, 21, 40
Birkhoff's theorem 144
Born–Infeld (–Plebański) nonlinear electrodynamics 73, 97, 145

Callaway, J. 52, 61, 139
Cartan, E. 14, 15
Christoffel bracket 10, 16, 19, 46, 63, 76, 77, 84, 111, 113, 124
charged test particle 92, 144
choice of structure group 18
criterion of validity (of a theory) 6
connection form 43
cosmic electromagnetic field 121
cosmological consequences of the field theory 17
Coulomb force 60, 69, 73, 94, 95–7, 139, 140, 142
covariance (principle of) 6, 21, 135
covariant differentiation 41, 43, 45
curvature form 43

D'Alembertian 30
de Sitter cosmological constant 114, 116
dipole moment 59
dipoles 57, 58
Dirac electrodynamics 97, 128, 131, 133
domain
  of applicability (of a theory) 5, 6, 13
  of verifiability (of a theory) 5

Einstein and Infeld 55
Einstein–Infeld–Hoffman technique 15, 16, 52, 54, 59, 61, 68
Einstein–Kaufman parameters 16

Einstein–Maxwell theory 6, 11, 60, 97, 136
Einstein and Straus theory 1, 14, 16, 31
Einstein tensor 9, 32, 37, 115
electric charge in electrostatic units 88
electromagnetic field with a cut-off 99
electromagnetic vector potential 14, 94, 131
empirical validity (test of) 1
energy momentum tensor 9, 11, 12, 13, 36, 37, 39, 54, 103, 113, 120, 135, 140, 143
equations of motion 11, 15, 59
equivalence of observers 1

Fermi coordinates 59
fiber 44
fiber metric 45
flat-space approximation 129
form invariance 84
frame components 43
Friedmann's cosmological principle 120

Gauss' theorem 56
general linear group 43
generalised torsion 49
geodesic deviation 32
geometrisation (strong principle of) 9, 14, 80
geometrisation of physics 2, 6, 10
Gregory, L. J. 68, 76, 90

Hermitian symmetry (principle of) 25, 29, 40, 45, 49, 50, 52, 138
Hermitian symmetry and charge conjugation 14, 16, 17, 21, 22–5, 49, 52, 146
Hermitian (fundamental) variables 24, 32
Hermitian variational principle 27
Hlavaty, V. 14, 40, 76, 79
Hubble expansion 12, 103–6

Hubble horizon 107

identification of the electromagnetic field 15
impossibility of a local test of the field theory 100
inertial mass 9
Infeld, L. 52, 61

Killing equations 81, 83, 108, 109
Killing vectors 109, 112
(Schrödinger–) Klein–Gordon equation 128, 134

Lie algebra 43
Lie bracket 43
Lie derivative 84
local coordinates 44
Lorentz force 16, 30, 52, 69, 75, 94
Lorentz group 2, 11, 43, 45, 48, 69, 84, 124

Mach's principle and the electric charge 121
McInnes, B. T. 17
Maxwell's energy–stress–momentum tensor 11
minimum coupling hypothesis 128–30, 132
metric hypothesis 38, 41, 46, 50–3, 80, 85, 131, 133, 138
metric tensor 28, 37, 45, 46, 85
Mie's theory 72
Minkowski inner product 43
(mathematical) model of reality 4, 7, 8
monopole solution 57

Newtonian approximation 59–61, 65
nonexistence of magnetic monopoles 16, 97
non-Riemannian geometry 16
nonsymmetric metric 14
nonverifiability of general relativity 135

Occam's razor (principle of simplicity) 39, 41
Ohm's law 15

PCT theorem 22
Papapetrou's solution 69, 73, 84, 95, 139
paracompact manifolds 20
parallel transfer 10, 19, 41
Pauli spin matrices 123

physics and philosophy 7
Poincaré frame 48
Poincaré group 43–5
Poisson equation 58, 64
Popper, K. 5
Popper's criterion of falsifiability 6
principle of equivalence 2, 11, 18, 21, 39, 103, 135
projection mapping 44

quantum mechanics and macrophysics 3, 26, 122, 145
quasi-stationary expansion 54, 75

Radford, C. J. 17, 43
radius of the universe 99
reducibility (principle of) 21, 41
Reissner–Nordström solution 12, 97, 136, 142
Ricci coefficients of rotation 44, 45, 46
Ricci tensor 9, 15, 20, 23, 24, 25, 28, 29, 31, 55, 84, 87, 95
Riemann–Christoffel tensor 20, 47, 49, 84, 108
Riemannian metric 9, 10, 11, 16, 26, 50, 51, 55, 81, 95, 124
Robertson–Walker metric 110, 111, 115
Russell, G. K. 15, 16, 31, 62, 68

Schrödinger's connection 16, 21, 29, 30, 40, 46, 71, 79, 87, 127
Schwarzschild singularity 100
Schwarzschild solution 9, 57, 85, 96, 97, 98, 99, 105, 121
screening of the electromagnetic field 10
semi-direct product 43
Stokes' theorem 56
strong field equations 14, 15, 30, 61
symmetry groups 80
symplectic group 48
Szekeres–Kruskal coordinates 104

tangent affine space 43
tangent bundle 45
Tonnelat, M -A. 39, 76–9, 86
torsion form 44
Treder, H. 15, 66, 67, 68, 141

universal symmetry 84–6

variational action principle 39, 51
variational derivation of the field equations 14

variational parameters 26–9
van der Waerden and Infeld 125

weak field approximation 30
weak field equations 14, 15, 29, 30, 46, 50,
53, 54, 61, 64, 86, 145
weak principle of equivalence 9
weak principle of geometrisation 17, 21,
38–41, 48–50, 100
Weyl, H. 32